FUTURE MEDIA DEVELOPMENT

PROFESSOR SANJAY ROUT

Copyright © Professor Sanjay Rout
All Rights Reserved.

This book has been published with all efforts taken to make the material error-free after the consent of the author. However, the author and the publisher do not assume and hereby disclaim any liability to any party for any loss, damage, or disruption caused by errors or omissions, whether such errors or omissions result from negligence, accident, or any other cause.

While every effort has been made to avoid any mistake or omission, this publication is being sold on the condition and understanding that neither the author nor the publishers or printers would be liable in any manner to any person by reason of any mistake or omission in this publication or for any action taken or omitted to be taken or advice rendered or accepted on the basis of this work. For any defect in printing or binding the publishers will be liable only to replace the defective copy by another copy of this work then available.

The Book Is Dedicated To All My Friends, Family, Parents And Almighty. Special Thanks To All the Reviewers, Designers and Technical Teams. For Whom This Entire Book Can Be Possible.

Contents

Foreword vii

Preface ix

Acknowledgements xi

Prologue xiii

    1. Introduction 1

    2. Chapter-1 5

    3. Chapter-2 8

    4. Chapter-3 44

    5. Chapter-4 55

    6. Chapter-5 62

    7. Chapter-6 70

    8. Chapter-7 79

References 97

About Author 99

About Publisher 101

Foreword

The book is written by Professor Sanjay Rout and Edited by Professor Prangyan Biswal Published by ISL Publications, India .All rights reserved. Any unauthorized reprint or use of this material is prohibited. No part of this book may be reproduced or transmitted in any form or by any means, electronic or mechanical, including photocopying, recording, or by any information storage and retrieval system without express written permission from the author/publisher. Please do not participate in or encourage piracy of copyrighted materials in violation of the author's rights. Purchase only authorized editions.

Preface

The book depicts all about current and future modern topics of development. This is an approach and perception of transformation in development. The book is for all cater to the audience throughout the globe.

Acknowledgements

I record deep sense of gratitude for my respected all my global Mentor's, Friend and Innovators for all constant direction, helpful discussion and valuable suggestions for writing this book. Due to his valuable suggestions and regular encouragement. I would be able to complete this work and fulfillment of my dream. All my global friends helped me enough during the entire project period like a torch in pitch darkness. I shall remain highly indebted to all throughout my life.

I acknowledge my deepest sense of gratitude to my learned parents, who has been throughout a source of Inspiration to me in conducting the study. Who helped me at various stages of the study directly or indirectly. He also enlightened me to follow the path of duty.

Special thanks to my son and spouse and almighty for their support in my work.

Prologue

Sl.No	Chapter	Page No
1	Introduction	6-9
2	Chapter-I	10-17
3	Chapter-II	18-62
4	Chapter-III	63-73
5	Chapter-IV	74-81
6	Chapter-V	82-88
7	Chapter-VI	89-97
8	Chapter-VII	98-117
9	References	118-118

౮౩

I
Introduction

The time period media, that's the plural of medium, refers to the communication channels thru which we disseminate news, tune, films, schooling, promotional messages and other records. It includes physical and online newspapers and magazines, tv, radio, billboards, telephone, the Internet, fax and billboards.

It describes the diverse approaches through which we speak in society. Because it refers to all method of verbal exchange, the whole thing ranging from a smartphone name to the evening news on tv can be referred to as media.

When speaking about reaching a completely large variety of humans we say mass media. Local media refers to, as an example, your neighborhood newspaper, or nearby/local TV/radio channels.

MediaWe used to get all our information and leisure via TV, radio, newspapers and magazines. Today the Internet is regularly taking up. Print newspapers are suffering as masses of millions of human beings every yr transfer to information resources on-line.

Different forms of media

Media can be broken down into main categories: broadcast and print. The Internet has also emerged as a primary participant, as a hastily-growing range of human beings globally get their information, movies, and so forth. Online.

Print Media includes all sorts of courses, which include newspapers, journals, magazines, books and reviews. It is the oldest type, and no matter struggling since the emergence of the Internet, continues to be utilized by a major percentage of the population.

Broadcast Media refers to radio and TV, which came onto the scene at the beginning and middle of the twentieth century respectively. Most people still get their information from TV and radio proclaims – but, experts expect that it's going to not be long before on line assets take over.

Over the beyond two decades, cable information has grown in importance.

The Internet – mainly websites and blogs – are hastily rising as viable and major channels of conversation as more and more human beings are seeking for information, amusement and educational cloth on line. The time period 'feasible,' in business, manner able to producing earnings for many years.

Virtually each a part of the Internet has grow to be a medium of communiqué – most unfastened email services have little bins that show advertisements and other messages.

The Internet as we realize it these days did now not certainly take off till the Nineties. In 1995, just 1% of the sector's population become online, compared to over forty nine% today. The belief of the Internet commenced in the 1960s inside the USA for the duration of the Cold War, whilst the navy and scientists have been concerned about a missile assault, which could knock out the smartphone system.

Stephen Hawking, a British theoretical physicist, cosmologist, writer and Director of Research on the Centre for Theoretical Cosmology in the University of Cambridge, once stated: "The media want superheroes in science simply as in every sphere of lifestyles, but there's surely a continuous variety of skills with out a clear dividing line."

What is social media?

Social media is a collective of online communication channels wherein communities engage, share content material and collaborate.

Websites and apps devoted to social networking, microblogging, forums, social bookmarking, wikis and social curation are examples of a few styles of social media.

The maximum famous social networking agencies are Facebook, Twitter, Google+ and Instagram.

Internet GrowthJust twenty-one years in the past, only a few humans the world over knew what the Internet turned into. Today it has turn out to be a part of our lives. It is destined to grow to be the number 1 channel for communicating with the sector's population. (Data Source: internetworldstats.Com)

Media 'is' or 'are'?

If media is the plural of medium, then one would suppose that it must be used grammatically in the plural – the verb that follows it ought to be in the plural form, shouldn't it?

However, in most literature it is used as a singular noun, and is interpreted as a collective singular, just like different collective nouns inclusive of 'team' or 'organization'. Therefore, to jot down the 'media is' is flawlessly proper nowadays. Some human beings may additionally insist it is incorrect, however it's far nevertheless suitable – languages are continuously evolving.

According to Collins Dictionary, media is:

"The way of communication that reach huge numbers of people, along with tv, newspapers, and radio."

It all started heaps of years ago

Human communique through designed channels – not via speech or gestures – dates back to many tens of lots of years ago while our ancient ancestors painted at the walls of caves.

The cave paintings at Lascaux in southwestern France, anticipated to be over 17,000 years vintage, are not any less feasible expressions of media than our modern-day TV indicates and magazines.

The Persian Empire – c. 550–330 BC – performed a first-rate role inside the records of human verbal exchange thru designed channels. Persian Emperor Cyrus the Great (c. 550 BC) advanced the primary ever real postal machine. It become an powerful intelligence-gathering equipment, known as Angariae, a time period that later indicated a tax device.

Thomas Carlyle (1795-1881), a Scottish truth seeker, satirical creator, essayist, historian and trainer claimed within the 1830s that the printing press created the contemporary world by using destroying feudalism. Many historians say that the arrival of the printing press turned into the start of what we recognise these days as media.

The time period media in its cutting-edge application referring to channels of communications become first utilized by Marshall McLuhan (1911-1980), a Canadian professor, philosopher, and public highbrow who said: "The media aren't toys; they have to no longer be inside the arms of Mother Goose and Peter Pan executives. They can be entrusted simplest to new artists, because they're art bureaucracy."

By the mid-1960s, the time period spread to trendy use within the United States, Canada and the United Kingdom.

FUTURE MEDIA DEVELOPMENT

II
Chapter-1

Media

Media plays an critical position in our day – to – day life. We stay up to date with the news and the matters that take place inside the international with the assist of it. One can know what's occurring inside the United States of America while he/she lives in India. It can make or damage the recognition of a person, corporation, political party, and many others. Let's find out extra approximately it.

Understanding Media

Media means the primary way of mass verbal exchange (broadcasting, publishing, and the Internet) regarded together. It has diverse forms which includes print, television, radio, and the net. The print is the oldest form of it. Whereas, the latest form of it's far the net.

Media

Print media has two factors i.E. Newspapers and magazines. It may also or may not be encouraged by way of effective people, political events, groups, and many others. This may additionally result in a biased insurance of that precise political birthday party, enterprise, or person.

Role of Media

Media is said to be the fourth pillar of democracy. One of its critical roles is to inform humans about the things which can be happening. It is likewise critical to present valid and actual information and allows to progress the society. It has additionally helped to get justice for a variety of instances.

For instance, The Ayushi Talvaar murder case, The Nirbhaya Rape case, Jessica Lal homicide case, and so forth. It additionally exposes numerous corrupt humans. Some different critical functions include influencing public opinion, determining the political time table, imparting a hyperlink between the government and the people, acting as a central authority watchdog, and affecting socialization.

Media and Big Business Houses

Big Business Houses and Media are interrelated. The media indirectly connects the common human beings to large business houses. These large enterprise homes pay the media company to put it on the market their products and services. Their classified ads are then proven on television, newspaper, radio, etc.

People get aware of the goods and services provided by using a agency and purchase their merchandise which advantage the corporations. Companies also gain a reputation due to this. This dating can be visible as a dating between an advertiser and a maker.

Media is the plural of medium (middle, center, middleman).

So Media is an intermediary with a view to shipping facts used to speak (like the press, radio, tv). Medias permit to disseminate records to a huge number of humans without customization of the message. That is why we talk also of mass media. However, the time period is utilized in more standard senses, to designate the manner of communication together with language, writing or tune.

What is media Technology?

That's Technology and methods that support human communication over distances in time and space.

In painting and humanities, it uses the medium time period to explain a technique used (example: watercolor is a medium).

• In Audio, video and computer, media help is physically saved content (inside the case of a unmarried document) or transferred content (inside the case of a message), track, movie, photographs or greater typically of records.

Other today various media:

• Newspapers,

• Radio,
• TV broadcasting video montages commented,
• Internet,

- Cinema.

Now that's we have some example of what may be these days media technologies, we are able to look to what's going to be the subsequent media technology. Here is some video and hyperlink in order you get in touch with those new media technologies:

- Mobile media offerings : Mobile goes to be the new mass media via definition.
- ePaper : will surely update the traditional newspaper within the future
- Wearables : your tomorrows garments
- Tangible Interfaces : new manner to apply your desk, read, play...
- I/O Brush : an other way to color

What is media era as an educational subject?

Media Technology is an engineering technology. As in all engineering technology, in media generation one in all the biggest challenges is to create value-effective solutions to realistic trouble via making use of scientific expertise to building matters inside the carrier of mankind.

Today Media technology are construct via electronic and pc structures but because the definition said it's a medium that's why operating in media technology can make you study loads about different field as :

- Physics (mechanics, optics)
- Chemistry
- Paper technology, substances generation
- Electronics, telecommunications
- Mathematics
- Computer science, human-gadget interplay
- Psychology, cognition science, belief psychology
- Communication science
- Sociology
- Economics, enterprise improvement, law
- Esthetics and layout

III
Chapter-2

Media History

Mass media refers to a numerous array of media technology that attain a massive target audience thru mass verbal exchange. The technology thru which this communication takes area consist of a diffusion of shops.

Broadcast media transmit facts electronically thru media inclusive of movies, radio, recorded tune, or tv. Digital media incorporates both Internet and cell mass communique. Internet media contain such services as e-mail, social media sites, websites, and Internet-based totally radio and television. Many other mass media shops have a further presence on the net, by using such means as linking to or running TV ads on-line, or dispensing QR Codes in outside or print media to direct cell customers to a internet site. In this manner, they can use the smooth accessibility and outreach competencies the Internet provides, as thereby without problems broadcast records at some point of many unique regions of the world concurrently and value-effectively. Outdoor media transmit facts thru such media as AR advertising; billboards; blimps; flying billboards (symptoms in tow of airplanes); placards or kiosks placed inside and outside buses, business buildings, shops, sports stadiums, subway cars, or trains; symptoms; or skywriting.[1] Print media transmit statistics thru physical gadgets, which includes books, comics, magazines, newspapers, or pamphlets.[2] Event organizing and public speakme also can be taken into consideration styles of mass media.[3]

The businesses that manipulate these technology, including film studios, publishing organizations, and radio and television stations, are also referred tó as the mass medía.

Íssúes wíth defínítión

Ín the past dúe twentíeth centúry, mass medía may be labeled íntó 8 mass medía índústríes: bóóks, the Ínternet, magazínes, móvies, newspapers, radíó, recórdíngs, and tv. The explósión óf vírtúal verbal exchange technólógy withín the late 20th and early 21st centúríes made óútstanding the qúery: what fórms óf medía shóúld be categórised as "mass medía"? Fór example, ít's far debatable whether tó cónsíst óf cell telephónes, cómpúter games (tógether with MMÓRPGs), and vídeó vídeó games withín the definítión. Ín the 2000s, a categóry called the "seven mass medía" became pópúlar.[cítatíón needed] Ín órder óf intródúctíón, they may be:

Prínt (bóóks, pamphlets, newspapers, magazínes, etc.) fróm the late fífteenth centúry

Recórdíngs (gramóphóne facts, magnetíc tapes, cassettes, cartrídges, CDs, and DVDs) fróm the past dúe nineteenth centúry

Cínema fróm appróximately 1900

Radíó fróm appróximately 1910

Televísión fróm abóút 1950

Ínternet fróm abóút 1990

Móbíle phónes fróm appróximately 2000

Each mass medíúm has íts persónal cóntent sórts, ínnóvatíve artísts, technícíans, and cómmercíal enterpríse módels. Fór ínstance, the Ínternet inclúdes blógs, pódcasts, ínternet web sítes, and númeróús óther technólógies búilt atóp the general dístríbútíón cómmúnity. The 6th and 7th medía, Ínternet and cell phónes, are regularly cíted tógether as dígital medía; and the fóúrth and 5th, radíó and TV, as bróadcast medía. Sóme argúe that vídeó games have evólved ríght íntó a dístínct mass shape óf medía.[6]

Whíle a cellphóne ís a twó-manner cónversatíón tóól, mass medía cómmúnícates tó a húge órganízatíón. Ín addítíón, the phóne has cónverted íntó a cell telephóne that's prepared wíth Ínternet access. A qúestíón aríses whether thís makes cell phónes a mass medíúm ór súrely a tóól úsed tó get ríght óf entry tó a mass medíúm (the Ínternet). There ís cúrrently a system wíth the aíd óf whích marketers and advertísers are able tó tap íntó satellítes, and bróadcast classífied ads and classífíed ads dírectly tó cellúlar telephónes, únsólícíted vía the smartphóne's cónsúmer.[cítatíón needed] Thís transmíssión óf mass advertísíng and marketing tó húndreds óf thóúsands óf húmans ís every óther fórm óf mass cómmúnícatíón.

Vídeó vídeó games wíll alsó be evólvíng ríght íntó a mass medíúm. Vídeó games (fór ínstance húgely múltíplayer ón líne fúnctíón-playing games (MMÓRPGs), alóng wíth RúneScape) próvíde a nót únúsúal gaming experíence tó míllíóns óf úsers thróúghóút the glóbe and pródúce the same messages and ideólógíes tó all their cústómers. Úsers every nów and then própórtíón the experience wíth each óther by way óf gamblíng ónlíne. Exclúdíng the Ínternet bút, ít's míles qúestíónable whether gamers óf vídeó vídeó games are sharing a cómmónplace revel ín when they play the game ín my ópíníón. Ít is víable tó speak abóút ín amazing detail the actívítíes óf a vídeó game wíth a chúm óne has by nó means perfórmed wíth, becaúse the revel ín is ídentícal tó each. The qúestíón, then, is whether that is a shape óf mass cómmúníqúe.[citatíón needed]

Characterístics

Fíve traits óf mass cómmúnícatíón were díagnósed by sócíólógíst Jóhn Thómpsón óf Cambrídge Ún̄íversíty:[7]

"[C]ómpríses each technical and instítútíónal strategies óf pródúctíón and dístríbútíón" – Thís is óbvíóús all thróúgh the recórds óf mass medía, fróm prínt tó the Ínternet, each apprópríate fór cómmercíal applícatíón

Ínvólves the "cómmódíficatíón óf symbólic paperwórk" – as the manúfactúríng óf súbstances is predícated ón its pótentíal tó manúfactúre and prómóte húge pórtíóns óf the paíntings; as radíó statíóns rely ón their tíme bóúght tó classífíed ads, só tóó newspapers rely ón their space fór the ídentícal reasóns

"[S]eparate cóntexts amóng the pródúctíón and receptíón óf data"

Íts "reach tó thóse 'a ways remóved' ín tíme and space, ín cómparísón tó the pródúcers"

"[Í]nfórmatíón dístríbútíón" – a "óne tó many" fórm óf cónversatíón, whereby merchandíse are heavíly pródúced and dísseminated tó a nótable qúantíty óf aúdíences

Mass vs. Maínstream and alternatíve

The tíme períód "mass medía" is ón óccasíón erróneóúsly úsed as a synónym fór "maínstream medía". Maínstream medía are óútstandíng fróm alternatíve medía by úsíng their cóntent and factór óf víew. Alternatíve medía alsó are "mass medía" óútlets wíthín the experience that they úse generatíón able tó reachíng many húman beíngs, despíte the fact that the target aúdíence is regúlarly smaller than the maínstream.

Ín cómmónplace úsage, the tíme períód "mass" denótes nót that a gíven númber óf peóple receíves the pródúcts, bút as a súbstítúte that the pródúcts are to be had in precept to a plurality of recipients.[7]

Forms of mass media

Broadcast

A family paying attention to a crystal radio in the Twenties.

Main articles: Radio broadcasting and Television

The sequencing of content in a broadcast is known as a agenda. With all technological endeavours some of technical terms and slang have evolved. Please see the listing of broadcasting terms for a thesaurus of terms used.

Radio and tv applications are allotted over frequency bands which might be relatively regulated inside the United States. Such law includes willpower of the width of the bands, variety, licensing, types of receivers and transmitters used, and appropriate content.

Cable television packages are regularly broadcast concurrently with radio and television applications, however have a extra restrained audience. By coding alerts and requiring a cable converter container at individual recipients' locations, cable also allows subscription-primarily based channels and pay-in line with-view services.

A broadcasting organization can also broadcast several packages simultaneously, via several channels (frequencies), as an example BBC One and Two. On the opposite hand, two or more establishments might also share a channel and each use it all through a hard and fast a part of the day, which includes the Cartoon Network/Adult Swim. Digital radio and digital tv may additionally transmit multiplexed programming, with numerous channels compressed into one ensemble.

When broadcasting is completed through the Internet the term webcasting is often used. In 2004, a new phenomenon occurred while a number of technologies mixed to provide podcasting. Podcasting is an asynchronous broadcast/narrowcast medium. Adam Curry and his associates, the Podshow, are main proponents of podcasting.

Film

Main article: Film

The term 'movie' encompasses motion photos as person tasks, in addition to the sphere in wellknown. The call comes from the photographic movie (also known as filmstock), traditionally the number one medium for recording and displaying motion images. Many different terms for film exist, together with motion snap shots (or just photos and "picture"), the silver display, photoplays, the cinema, photograph indicates, flicks, and maximum not unusual, films.

Films are produced by using recording people and objects with cameras, or with the aid of creating them the usage of animation strategies or computer graphics. Films contain a chain of man or woman frames, but when these pics are shown in speedy succession, an phantasm of motion is created. Flickering between frames isn't visible because of an effect known as persistence of vision, wherein the eye keeps a visible photograph for a fraction of a 2nd after the source has been removed. Also of relevance is what reasons the belief of motion: a psychological effect identified as beta movement.

Film is taken into consideration by using many[who?] to be an vital artwork form; films entertain, educate, enlighten, and encourage audiences. Any movie can become a global appeal, mainly with the addition of dubbing or subtitles that translate the film message. Films are also artifacts created by using particular cultures, which replicate those cultures, and, in turn, have an effect on them.[who?]

Video video games

Shopping carts for children geared up with gaming computer systems.

A video game is a pc-controlled game wherein a video display, including a display or television, is the primary remarks tool. The term "pc recreation" additionally includes video games which display only textual content (and that may, consequently, theoretically be performed on a teletypewriter) or which use different strategies, which include sound or vibration, as their primary remarks tool, however there are very few new games in these classes.[who?] There always have to also be a few sort of input device, commonly inside the form of button/joystick combinations (on arcade video games), a keyboard and mouse/trackball combination (computer games), a controller (console games), or a mixture of any of the above. Also, more esoteric gadgets had been used for input, e.G., the participant's movement. Usually there are regulations and dreams, but in more open-ended games the participant can be unfastened to do anything they like within the confines of the digital universe.

In not unusual usage, an "arcade game" refers to a recreation designed to be played in an established order in which purchasers pay to play on a consistent with-use basis. A "computer sport" or "PC game" refers to a recreation that is performed on a non-public laptop. A "Console game" refers to at least one that is played on a tool specially designed for using such, whilst interfacing with a trendy tv set. A "online game" (or "videogame") has advanced into a catchall phrase that encompasses the aforementioned in conjunction with any game made for any other tool, such as, however no longer confined to, advanced calculators, cellular phones, PDAs, etc.

Audio recording and replica

Sound recording and replica is the electric or mechanical re-introduction or amplification of sound, often as music. This entails the usage of audio device along with microphones, recording devices, and loudspeakers. From early beginnings with the invention of the phonograph the use of only mechanical strategies, the sector has superior with the invention of electrical recording, the mass manufacturing of the 78 record, the magnetic wire recorder observed via the tape recorder, the vinyl LP report. The invention of the compact cassette inside the Sixties, followed by means of Sony's Walkman, gave a primary improve to the mass distribution of track recordings, and the discovery of digital recording and the compact disc in 1983 introduced large improvements in ruggedness and great. The most latest traits have been in digital audio players.

An album is a set of related audio recordings, released together to the public, commonly commercially.

The time period report album originated from the truth that seventy eight RPM Phonograph disc statistics were saved together in a e book similar to a photograph album. The first collection of facts to be called an "album" changed into Tchaikovsky's Nutcracker Suite, launch in April 1909 as a 4-disc set via Odeon information.[8][9] It retailed for 16 shillings – about £15 in current forex.

A song video (also promo) is a brief film or video that accompanies a whole piece of song, most commonly a track. Modern tune videos had been frequently made and used as a advertising tool intended to sell the sale of tune recordings. Although the origins of music movies cross lower back much further, they came into their very own in the Nineteen Eighties, while Music Television's format became primarily based on them. In the 1980s, the time period "rock video" was often used to explain this shape of enjoyment, despite the fact that the term has fallen into disuse.

Music movies can accommodate all styles of filmmaking, along with animation, live action films, documentaries, and non-narrative, summary film.

Internet

See additionally: Digital media

The Internet (also regarded simply as "the Net" or much less precisely as "the Web") is a extra interactive medium of mass media, and can be briefly described as "a network of networks". Specifically, it's far the global, publicly reachable community of interconnected pc networks that transmit statistics with the aid of packet switching the use of the same old Internet Protocol (IP). It includes thousands and thousands of smaller home, instructional, enterprise, and governmental networks, which together convey numerous records and services, including email, on line chat, record switch, and the interlinked net pages and different documents of the World Wide Web.

Contrary to some commonplace usage, the Internet and the World Wide Web aren't synonymous: the Internet is the machine of interconnected pc networks, related through copper wires, fiber-optic cables, wi-fi connections and so on.; the Web is the contents, or the interconnected files, related by links and URLs. The World Wide Web is obtainable through the Internet, in conjunction with many different services which include email, report sharing and others defined under.

Toward the give up of the 20th century, the arrival of the World Wide Web marked the primary era in which maximum people ought to have a means of publicity on a scale similar to that of mass media. Anyone with an internet site has the potential to cope with a global audience, even though serving to high degrees of web traffic is still exceptionally luxurious. It is feasible that the rise of peer-to-peer technologies may also have started the procedure of making the value of bandwidth manageable. Although a extensive quantity of facts, imagery, and remark (i.E. "content") has been made available, it's far frequently tough to decide the authenticity and reliability of data contained in web pages (in many instances, self-posted). The invention of the Internet has also allowed breaking news tales to attain around the world within mins. This rapid growth of instantaneous, decentralized verbal exchange is regularly deemed probable to trade mass media and its dating to society.

"Cross-media" manner the concept of distributing the same message via specific media channels. A similar concept is expressed inside the information enterprise as "convergence". Many authors recognize go-media publishing to be the ability to put up in both print and on the net with out guide conversion effort. An increasing quantity of wi-fi devices with at the same time incompatible data and display formats make it even extra difficult to reap the objective "create once, put up many".

The Internet is speedy turning into the middle of mass media. Everything is turning into accessible through the internet. Rather than picking up a newspaper, ór watching the ten ó'clóck ínfórmatión, peóple can lóg óntó the ínternet tó get the ínfórmatión they need, when they need ít. Fór instance, many emplóyees cóncentrate tó the radíó thrú the Ínternet at the same tíme as síttíng at their desk.

Even the training device ís predícated ón the Ínternet. Teachers can tóúch the cómplete class by way óf sendíng óne e maíl. They may addítíónally have ínternet pages ón whích cóllege stúdents can get any óther replíca óf the elegance óútlíne ór assignments. Sóme classes have class blógs ín which stúdents are reqúíred tó póst weekly, wíth stúdents graded ón their cóntríbútíóns.

Blógs (net lógs)

Blóggíng, tóó, has emerge as a pervasíve fórm óf medía. A blóg ís a ínternet site, úsúally maintained by way óf an persón, with órdinary entríes óf remark, descríptíóns óf óccasíóns, ór interactíve medía alóng wíth píx ór vídeó. Entríes are úsúally displayed ín óppósíte chrónólógícal órder, wíth maxímúm cúrrent pósts próven ón pinnacle. Many blógs óffer remark ór news ón a selected challenge; óthers feature as greater persónal ón-líne díaries. A tradítíónal blóg cómbínes text, ímages and óther píx, and links tó óther blógs, ínternet pages, and assócíated medía. The capacity fór readers tó gó away remarks ín an ínteractíve layóút ís an crúcíal part óf many blógs. Móst blógs are prímaríly textúal, althóúgh sóme cógnízance ón artwórk (artlóg), images (phótóblóg), sketchblóg, vídeós (vlóg), track (MP3 weblóg), aúdíó (pódcastíng) are a part óf a múch bróader cómmúnity óf sócíal medía. Micróblóggíng ís anóther kínd óf blóggíng whích cónsists óf blógs wíth very qúíck pósts.

RSS feeds

RSS ís a fórmat fór syndícatíng ínfórmatíón and the cóntent óf ínfórmatíón-líke sítes, inclúsive óf móst impórtant news sítes líke Wíred, news-óríentated cómmúníty websites líke Slashdót, and prívate blógs. Ít ís a famíly óf Web feed fórmats úsed tó públísh regúlarly úp tó date cóntent súch as weblóg entríes, news headlínes, and pódcasts. An RSS recórd (that's called a "feed" ór "ínternet feed" ór "channel") inclúdes either a súmmary óf cóntent fróm an assócíated ínternet web page ór the óverall text. RSS makes ít feasíble fór húman beíngs tó hóld úp wíth ínternet sítes ín an aútómatíc way that may be píped íntó úníqúe prógrams ór fíltered presentatíóns.

Pódcast

Maín artícle: Pódcast

A pódcast ís a seqúence óf vírtúal-medía fíles whích are allótted óver the Ínternet the úse óf syndícatíón feeds fór playback ón pórtable medía players and cómpúter systems. The term pódcast, líke bróadcast, can refer eíther tó the cóllectíón óf cóntent ítself ór tó the techníque thróúgh whích ít's míles syndícated; the latter ís alsó referred tó as pódcastíng. The hóst ór aúthór óf a pódcast ís regúlarly called a pódcaster.

Móbíle

Maín artícle: Móbíle medía

Móbíle phónes had been added ín Japan ín 1979 hówever have becóme a mass medía móst effectíve ín 1998 when the fírst dównlóadable rínging tónes had been delívered ín Fínland. Sóón móst sórts óf medía cóntent had been íntródúced ón cell telephónes, tablets and óther transpórtable gadgets, and these days the entíre valúe óf medía cónsúmed ón móbíle húgely exceeds that óf net cóntent materíal, and became well wórth óver 31 bíllíón búcks ín 2007 (súpply Ínfórma). The cell medía cóntent materíal inclúdes óver 8 bíllíón greenbacks really wórth óf móbíle músíc (rínging tónes, ríngback tónes, trúetónes, MP3 fíles, karaóke, sóng móvíes, músíc streamíng ófferíngs and só ón.); óver fíve bíllíón búcks wórth óf cell gamíng; and díverse ínfórmatíón, enjóyment and advertísing and marketíng ófferíngs. Ín Japan móbíle smartphóne bóóks are só famóús that fíve óf the 10 qúalíty-sellíng revealed bóóks were óríginally laúnched as cellúlar phóne bóóks.

Símílar tó the ínternet, móbíle ís alsó an ínteractíve medía, hówever has far wíder reach, wíth 3.3 bíllíón cellúlar cellphóne cústómers at the stóp óf 2007 tó óne.3 bíllíón net úsers (sóúrce ÍTÚ). Líke e maíl ón the net, the pínnacle applícatíón ón cell ís líkewíse a nón-públíc messaging carríer, bút SMS textúal cóntent messagíng ís útílízed by óver 2.4 bíllíón húman beíngs. Practícally all net servíces and applícatíóns exíst ór have cómparable cóúsíns ón móbíle, fróm search tó múltíplayer games tó dígital wórlds tó blógs. Móbíle has several precíse benefíts whích many cell medía púndíts claím make móbíle a móre effectíve medía than eíther TV ór the net, startíng wíth cell beíng permanently carríed and cónstantly cónnected. Móbíle has the fírst-class target market accúracy and ís the best mass medía wíth a íntegrated payment channel avaílable tó each úser wíthóút any credít scóre cards ór PayPal bílls ór even an age restríctíón. Móbíle ís freqúently referred tó as the seventh Mass Medíúm and bóth the fóúrth dísplay (íf cóúntíng cínema, TV and PC mónítórs) ór the 0.33 dísplay (cóúntíng best TV and PC).

Prínt medía

Main articles: Newspaper and Magazine
See additionally: Publishing § Industry sub-divisions, and Printing

Magazine

A mag is a periodical guide containing a variety of articles, generally financed by advertising and marketing or buy by using readers.

Magazines are generally posted weekly, biweekly, month-to-month, bimonthly or quarterly, with a date on the duvet that is earlier of the date it's miles sincerely posted. They are frequently printed in coloration on lined paper, and are bound with a tender cover.

Magazines fall into two huge categories: customer magazines and business magazines. In exercise, magazines are a subset of periodicals, wonderful from those periodicals produced by way of scientific, artistic, educational or unique interest publishers that are subscription-most effective, extra luxurious, narrowly limited in stream, and frequently have little or no advertising.

Magazines can be labeled as:

General interest magazines (e.G. Frontline, India Today, The Week, The Sunday Times and many others.)

Special interest magazines (women's, sports activities, enterprise, scuba diving, and so forth.)

Newspaper

A panel in the Newseum in Washington, D.C., showing newspaper headlines from the day after September 11.

A newspaper is a book containing information and facts and marketing, normally published on low-cost paper known as newsprint. It may be trendy or unique hobby, most customarily posted each day or weekly. The maximum essential function of newspapers is to inform the public of sizable events.[10] Local newspapers inform neighborhood groups and include advertisements from nearby agencies and services, whilst national newspapers generally tend to cognizance on a subject matter, which may be exampled with "The Wall Street Journal" as they provide news on finance and business associated-topics.[10] The first revealed newspaper turned into published in 1605, and the form has thrived even within the face of competition from technology together with radio and television. Recent trends at the Internet are posing fundamental threats to its commercial enterprise version, however. Paid move is declining in maximum nations, and advertising sales, which makes up the bulk of a newspaper's profits, is moving from print to on-line; a few commentators, nevertheless, point out that historically new media including radio and television did no longer entirely supplant current.

The net has challenged the click as an opportunity supply of records and opinion however has additionally furnished a brand new platform for newspaper corporations to attain new audiences.[11] According to the World Trends Report, among 2012 and 2016, print newspaper flow continued to fall in nearly all regions, aside from Asia and the Pacific, where the dramatic increase in income in a few pick out countries has offset falls in historically robust Asian markets along with Japan and the Republic of Korea. Most extensively, between 2012 and 2016, India's print movement grew by 89 in step with cent.[12]

Outdoor media

Political advertisements on a billboard inside the Netherlands in 2019.

Outdoor media is a form of mass media which contains billboards, signs and symptoms, placards located inside and outside business buildings/objects like stores/buses, flying billboards (signs in tow of airplanes), blimps, skywriting, AR Advertising. Many business advertisers use this form of mass media whilst advertising and marketing in sports activities stadiums. Tobacco and alcohol producers used billboards and different out of doors media substantially. However, in 1998, the Master Settlement Agreement between the United States and the tobacco industries prohibited the billboard advertising and marketing of cigarettes. In a 1994 Chicago-based study, Diana Hackbarth and her colleagues revealed how tobacco- and alcohol-based billboards have been concentrated in poor neighbourhoods. In other city centers, alcohol and tobacco billboards have been a whole lot greater focused in African-American neighborhoods than in white neighborhoods.[1]

Purposes

Mass media encompasses a whole lot extra than just information, even though it is every now and then misunderstood in this way. It may be used for various functions:

Advocacy, both for commercial enterprise and social issues. This can encompass advertising, advertising, propaganda, public relations, and political verbal exchange.

Entertainment, traditionally through performances of performing, tune, and TV indicates along with mild analyzing; because the past due twentieth century also via video and computer games.

Públic service búlletíns and emergency índicatórs (that can be úsed as pólítical device tó speak própaganda tó the públic).[13]

Prófessións related tó mass medía

Jóúrnalísm

Jóúrnalism ís the field óf gatheríng, stúdying, verifyíng and óffering statístics regardíng módern óccasións, trends, próblems and thóse. Thóse whó exercíse jóúrnalism are knówn as jóúrnalists.

News-óríentated jóúrnalism ís ón óccasión descríbed as the "fírst hard draft óf recórds" (attríbúted tó Phíl Graham), dúe tó the fact newshóúnds regúlarly file essentíal actívities, pródúcing ínformatión articles ón qúick time límits. While únder straín tó be fírst with their tales, news medía córpóratións úsúally edít and próofread their reviews príor tó ebóok, adhering tó each órganízatión's standards óf accúracy, best and fashíon. Many news cómpanies claim próud tradítions óf cónserving aúthorítíes óffícials and establíshments accóúntable tó the general públic, at the same time as medía crítics have raised qúestions ón cónserving the click itself respónsíble tó the standards óf próféssiónal jóúrnalism.

Públic famíly members

Públic relatións ís the art and science óf managing cómmúnicatión amóng an enterpríse and its key públics tó cónstrúct, cóntról and maintain its nice phótógraph. Examples cónsist óf:

Córpóratións úse advertising públic relatións tó cónvey ínformatión abóút the pródúcts they manúfactúre ór ófferings they óffer tó capacity clients tó aíd their direct sales effórts. Typically, they assist sales ínside the qúick and lóng term, setting úp and búrníshing the órganizatión's brandíng fór a stúrdy, óngóing marketplace.

Córpóratións alsó úse públic members óf the family as a vehicle tó attaín legíslatórs and óther pólíticíans, searching fór favórable tax, regúlatóry, and dífferent remedy, and they'll úse públic relatións tó painting themselves as enlightened emplóyers, ín gúide óf húman-resóúrces recrúíting packages.

Nónprófit cómpaníes, which inclúde facúlties and úniversities, hóspitals, and húman and sócial carrier órganizatións, úse públic relatións ín assist óf cógnízance applícatións, fúnd-raísing packages, team óf wórkers recrúíting, and tó íncrease patrónage óf their services.

Pólíticíans úse públic relatións tó attract vótes and raise móney, and while súccessfúl ón the póll cóntainer, tó prómóte and shield their service ín wórkplace, with an eye tó the súbseqúent electíon ór, at career's stóp, tó their legacy.

Publishing

Museum currator shows a child early printing methods.

Publishing is the enterprise concerned with the production of literature or statistics – the pastime of creating records available for public view. In a few instances, authors may be their personal publishers.

Traditionally, the time period refers to the distribution of published works inclusive of books and newspapers. With the appearance of virtual information structures and the Internet, the scope of publishing has improved to include web sites, blogs, etc.

As a enterprise, publishing consists of the improvement, advertising, production, and distribution of newspapers, magazines, books, literary works, musical works, software, different works dealing with records.

Publication is also critical as a criminal idea; (1) because the method of giving formal notice to the sector of a considerable intention, for example, to marry or input financial disaster, and; (2) because the crucial precondition of being capable of claim defamation; that is, the alleged libel should were posted.

Software publishing

A software program publisher is a publishing organisation in the software enterprise between the developer and the distributor. In some companies, two or all three of those roles can be mixed (and indeed, can also are living in a single person, specifically in the case of shareware).

Software publishers frequently license software program from developers with particular obstacles, which includes a time restriction or geographical area. The terms of licensing vary exceedingly, and are normally mystery.

Developers may additionally use publishers to reach large or overseas markets, or to avoid focussing on advertising. Or publishers might also use developers to create software to meet a market want that the publisher has diagnosed.

Internet Based Professions

A YouTuber is everyone who has made their reputation from growing and selling movies on the public video-sharing web site, YouTube. Many YouTube celebrities have made a profession from their website thru sponsorships, advertisements, product placement, and community help.

History

Early wooden printing press, depicted in 1520.

The hístóry óf mass medía can be traced back tó the tímes when dramas had been accómplíshed ín díverse histórìc cúltúres. Thís was the fírst tíme when a fórm óf medía becóme "bróadcast" tó a múch bróader aúdíence. The fírst dated prínted ebóók recógnísed ís the "Díamónd Sútra", revealed ín Chína ín 868 AD, even thóúgh ít ís apparent that bóóks were prínted ín advance. Móvable clay kínd changed íntó ínvented ín 1041 ín Chína. Hówever, becaúse óf the slów spread óf líteracy tó the lóads ín Chína, and the íncredíbly excessíve fee óf paper there, the earlíest prínted mass-medíúm túrned íntó próbably Eúrópean famóús prínts fróm appróxímately 1400. Althóúgh thóse had been pródúced ín massíve númbers, ónly a few early examples líve ón, ór even móst recógnízed tó be prínted befóre appróxímately 1600 have nót súrvíved. The tíme períód "mass medía" túrned íntó cóíned wíth the creatíón óf prínt medía, whích ís óútstandíng fór beíng the fírst ínstance óf mass medía, as we úse the term nówadays. Thís shape óf medía started ín Eúrópe ín the Míddle Ages.

Jóhannes Gútenberg's ínventíón óf the príntíng press allówed the mass manúfactúríng óf bóóks tó brúsh the kíngdóm. He públíshed the fírst ebóók, a Latín Bíble, ón a príntíng press wíth móvable type ín 1453. The ínventíón óf the príntíng press gave rise tó sóme óf the prímary types óf mass cómmúníqúe, with the aíd óf allówíng the e-bóók óf bóóks and newspapers ón a scale lóts larger than changed íntó fórmerly feasíble.[14][15][16] The ínventíón alsó cónverted the way the wórld receíved públíshed súbstances, despíte the fact that bóóks remaíned tóó híghly-príced absólútely tó be called a mass-medíúm fór at least a centúry after that. Newspapers evólved fróm appróxímately 1612, wíth the prímary ínstance ín Englísh ín 1620;[17] bút they tóók tíll the 19th centúry tó attaín a mass-aúdíence ímmedíately. The fírst hígh-stream newspapers aróse ín Lóndon wíthín the early 1800s, alóng wíth The Tímes, and had been made póssíble by the ínventíón óf excessíve-speed rótary steam príntíng presses, and raílróads whích allówed húge-scale dístríbútíón óver húge geógraphícal regíóns. The íncrease ín móve, bút, bróúght abóút a declíne ín cómments and ínteractívíty fróm the readershíp, makíng newspapers a greater óne-way medíúm.[18][19][20][21]

The wórd "the medía" started fór úse ín the 1920s.[22] The perceptíón óf "mass medía" was typícally límíted tó prínt medía úp tíll the súbmít-Secónd Wórld War, when radíó, tv and vídeó were delívered. The aúdíó-vísíble facílítíes have becóme very pópúlar, dúe tó the fact they fúrníshed each facts and entertaínment, dúe tó the fact the cólóúr and sóúnd engaged the vísítórs/lísteners and as ít was less díffícúlt fór móst óf the peóple tó passively watch TV or listen to the radio than to actively read. In recent instances, the Internet grow to be the latest and maximum famous mass medium. Information has end up with no trouble to be had through websites, and effortlessly available via engines like google. One can do many activities on the equal time, which includes gambling games, paying attention to tune, and social networking, no matter region. Whilst other types of mass media are confined in the form of facts they are able to provide, the internet contains a large percent of the sum of human information through such things as Google Books. Modern day mass media includes the internet, mobile telephones, blogs, podcasts and RSS feeds.[23]

During the twentieth century, the growth of mass media turned into driven through generation, including that which allowed a great deal duplication of cloth. Physical duplication technology together with printing, record urgent and film duplication allowed the duplication of books, newspapers and films at low fees to huge audiences. Radio and tv allowed the digital duplication of data for the first time. Mass media had the economics of linear replication: a single work may want to make money. An example of Riel and Neil's idea. Proportional to the number of copies bought, and as volumes went up, unit costs went down, growing earnings margins similarly. Vast fortunes had been to be made in mass media. In a democratic society, the media can serve the voters about issues regarding government and corporate entities (see Media impact). Some don't forget the awareness of media ownership to be a danger to democracy.[24]

Mergers and acquisitions

Between 1985 and 2018 approximately 76,720 deals were announced inside the media enterprise. This sums up to an normal price of round 5,634 billion USD.[25] There have been 3 major waves of M&A in the mass media zone (2000, 2007 and 2015), whilst the maximum active 12 months in phrases of numbers become 2007 with round three,808 offers. The United States is the maximum prominent united states of america in media M&A with 41 of the pinnacle 50 deals having an acquirer from the US.

The largest deal in records become the acquisition of Time Warner through AOL Inc. For 164,746.86 million USD.

Influence and sociology

Main article: Influence of mass media

This phase is written like a personal reflection, private essay, or argumentative essay that states a Wikipedia editor's personal emotions or presents an authentic argument approximately a subject. Please assist impróve ít vía rewríting ít ín an encyclópedíc style. (Febrúary 2013) (Learn hów and when tó díspóse óf thís template message)

Glóbe ícón.

The examples and angle ín thís segment may nót represent a wórldwíde víew óf the díffícúlty. Yóú may alsó enhance thís sectíón, speak the próblem ón the speak web page, ór create a brand new sectíón, as appróprìate. (March 2015) (Learn hów and whílst tó pút óff thís template message)

Límíted-resúlts ídea, at fírst examíned wíthín the 1940s and Fífties, cónsíders that becaúse peóple typícally select what medía tó have interactíón with based tótally ón what they already trúst, medía exerts a neglígíble have an effect ón. Class-dómínant príncíple argúes that the medía reflects and prójects the víew óf a minóríty elíte, whích cóntróls ít. Cúltúralíst theóry, whích changed íntó develóped in the Níneteen Eíghtíes and Níneties, cómbínes the óther twó theóríes and claíms that peóple engage with medía tó create their persónal meaníngs óút of the píx and messages they receíve. Thís cóncept states that aúdíence partícípants play an energétíc, as óppósed tó passíve fúnctíón in terms óf mass medía.

There ís an edítórial that argúes 90 percentage óf all mass medía whích inclúdes radíó bróadcast netwórks and prógramíng, vídeó ínfórmatíón, spórts leísúre, and óthers are ówned vía 6 main gróúps (GE, News-Córp, Dísney, Víacóm, Tíme Warner, and CBS).[26] Accórdíng tó Mórrís Creatíve Gróúp, these síx agencíes revamped 200 bíllíón greenbacks ín sales ín 2010. Móre range ís brewíng amóng many gróúps, hówever they have lately merged tó fórm an elíte that have the energy tó manípúlate the narratíve óf stóries and adjúst peóple's belíefs. Ín the new medía-púshed age we stay ín, advertísing and marketíng has móre valúe than ever earlíer than dúe tó the variíoús methóds ít ís able tó be carríed óút. Advertísements can persúade cítízens tó púrchase a specífic pródúct ór have cónsúmers avóíd a partícúlar pródúct. The defínítíón óf what ís acceptable with the aíd óf sócíety may be clósely díctated vía the medía ín regards tó the qúantíty óf ínterest ít gets.

The dócúmentary Súper Síze Me descríbes hów córpóratíóns líke McDónald's were súed ín the beyónd, the plaíntíffs claímíng that ít was the faúlt ín theír límínal and súblímínal advertísing that "cómpelled" them tó búy the pródúct. The Barbíe and Ken dólls óf the Níneteen Fíftíes are fróm tíme tó tíme referred tó as the main mótíve fór the óbsessíón ín cúrrent-day sócíety fór gírls tó be skínny and men tó be búff. After the attacks óf September 11, the medía gave massíve cóverage óf the óccasíón and úncóvered Ósama Bín Laden's gúílt fór the assaúlt, statístícs they have been informed through the government. This formed the public opinion to support the warfare on terrorism, and later, the conflict on Iraq. A foremost subject is that due to this intense strength of the mass media, portraying inaccurate records ought to lead to an enormous public concern. In his book The Commercialization of American Culture, Matthew P. McAllister says that "a well-evolved media system, informing and coaching its residents, helps democracy circulate in the direction of its perfect country."[1]

In 1997, J. R. Finnegan Jr. And K. Viswanath identified three foremost consequences or features of mass media:

The Knowledge Gap: The mass media affects knowledge gaps due to elements which includes "the volume to which the content material is attractive, the diploma to which facts channels are on hand and ideal, and the amount of social conflict and diversity there may be in a network".

Agenda Setting: People are encouraged in how they consider issues because of the selective nature of what media agencies select for public consumption. After publicly disclosing that he had prostate cancer previous to the 2000 New York senatorial election, Rudolph Giuliani, the mayor of New York City (aided through the media) sparked a big priority elevation of the most cancers in human beings's consciousness. This turned into due to the fact information media started out to file at the risks of prostate cancer, which in flip caused a more public focus about the disease and the need for screening. This capability for the media for you to change how the general public thinks and behaves has befell on different activities. In mid-Seventies when Betty Ford and Happy Rockefeller, wives of the then-President after which-Vice President respectively, had been each diagnosed with breast cancer. J. J. Davis states that "when risks are highlighted within the media, particularly in exquisite detail, the extent of time table putting is likely to be based on the degree to which a public feel of concern and hazard is provoked". When looking to set an schedule, framing may be invaluably useful to a mass media business enterprise. Framing includes "taking a management role in the agency of public discourse approximately an difficulty". The media is stimulated by way of the desire for balance in coverage, and the resulting pressures can come from corporations with particular political movement and advocacy positions. Finnegan and Viswanath say, "businesses, institutions, and advocates compete to become aware of problems, to move them onto the public time table, and to define the issues symbolically" (1997, p. 324).

Cultivation of Perceptions: The extent to which media exposure shapes audience perceptions through the years is called cultivation. Television is a not unusual enjoy, in particular in places like the United States, to the factor wherein it can be described as a "homogenising agent" (S. W. Littlejohn). However, in place of being merely a result of the TV, the impact is regularly based on socioeconomic factors. Having a prolonged exposure to TV or movie violence might affect a viewer to the quantity wherein they actively suppose network violence is a trouble, or alternatively discover it justifiable. The resulting notion is probably to be different depending on wherein people live but.[1]

Since the Nineteen Fifties, whilst cinema, radio and TV began to be the primary or the most effective source of facts for a larger and larger percentage of the populace, these media started to be considered as imperative contraptions of mass control.[27][28] Up to the point that it emerged the idea that after a rustic has reached a excessive level of industrialization, the usa itself "belongs to the individual that controls communications."[29]

Mass media play a tremendous position in shaping public perceptions on a spread of important problems, each through the information that is disbursed through them, and thru the interpretations they location upon this records.[27] They additionally play a massive position in shaping current lifestyle, with the aid of choosing and portraying a particular set of beliefs, values, and traditions (an entire manner of life), as reality. That is, by portraying a sure interpretation of truth, they form truth to be extra in step with that interpretation.[28] Mass media additionally play a important role in the spread of civil unrest sports together with anti-authorities demonstrations, riots, and trendy strikes.[30] That is, the use of radio and tv receivers has made the unrest influence among cities no longer most effective by using the geographic area of cities, but also by means of proximity in the mass media distribution networks.

In 2010, Americans should activate their tv and locate 24-hour news channels as well as music motion pictures, nature documentaries, and fact suggests approximately everything from hoarders to style fashions. That's not to say films to be had on demand from cable providers or tv and video to be had online for streaming or downloading. Half of U.S. Families obtain a day by day newspaper, and the common person holds 1.Nine magazine subscriptions (State of the Media, 2004) (Bilton, 2007). A University of California, San Diego have a look at claimed that U.S. Families consumed a tótal óf abóút 3.6 zettabytes óf recórds ín 2008—the dígital equál óf a 7-fóót hígh stack óf bóóks óverlayíng the entíre Úníted States—a 350 percentage grówth becaúse 1980 (Ramsey, 2009). Amerícans are úncóvered tó medía ín taxícabs and búses, ín schóól róóms and medícal dóctórs' wórkplaces, ón híghways, and ín aírplanes. We can start tó óríent óúrselves ín the ínfórmatíón clóúd thrú parsíng what róles the medía fílls ín sócíety, analyzíng íts recórds ín sócíety, and searchíng ón the way technólógícal ímpróvements have helped carry ús tó ín whích we're these days.

What Dóes Medía Dó fór Ús?

Medía fúlfílls númeróús fúndamental róles ín óúr sócíety. Óne apparent fúnctíón ís amúsement. Medía can act as a spríngbóard fór óúr ímagínatíóns, a sóúrce óf fable, and an óútlet fór escapísm. Ín the 19th centúry, Víctórían readers dísíllúsíóned by way óf the grímness óf the Índústríal Revólútíón fóúnd themselves drawn ínto fírst-rate wórlds óf faíríes and óther fíctítíóús beíngs. Ín the fírst decade óf the twenty fírst centúry, Amerícan televísíón víewers shóúld peek ín ón a cónflícted Texas excessíve cóllege sóccer gróúp ín Fríday Níght Líghts; the víólence-plagúed drúg exchange ín Baltímóre ín The Wíre; a Síxtíes-Manhattan advert enterpríse ín Mad Men; ór the últímate súrvívíng band óf húmans ín a far óff, míserable destíny ín Battlestar Galactíca. Thróúgh bríngíng ús testímóníes óf a wíde varíety, medía has the strength tó take ús away fróm óúrselves.

Medía can alsó próvíde ínfórmatíón and traíníng. Ínfórmatíón can cóme ín many búreaúcracy, and ít may ónce ín a whíle be díffícúlt tó separate fróm amúsement. Tóday, newspapers and ínfórmatíón-óríentated televísíón and radíó packages make avaílable memóríes fróm acróss the glóbe, allówíng readers ór vísítórs ín Lóndón tó access vóíces and mótíón píctúres fróm Baghdad, Tókyó, ór Búenós Aíres. Bóóks and magazínes óffer a better examíne a húge range óf súbjects. The lóóse ón-líne encyclópedía Wíkípedía has artícles ón súbjects fróm presídentíal nícknames tó tóddler pródígíes tó tóngúe twísters ín varíóús langúages. The Massachúsetts Ínstítúte óf Technólógy (MÍT) has públíshed free lectúre nótes, checks, and aúdíó and vídeó recórdíngs óf lessóns ón íts ÓpenCóúrseWare websíte, allówíng everyóne wíth an Ínternet cónnectíón get admíssíón tó tó wórld-elegance próféssórs.

Anóther benefícíal factór óf medía ís íts abílíty tó behave as a públíc díscússíón bóard fór the dialógúe óf essentíal íssúes. Ín newspapers ór dífferent períódícals, letters tó the edítór allów readers tó respónd tó newshounds or to voice their reviews at the issues of the day. These letters had been an crucial part of U.S. Newspapers even when the kingdom was a British colony, and that they have served as a means of public discourse ever since. The Internet is a basically democratic medium that permits anyone who can get online the capacity to express their critiques thru, for example, blogging or podcasting—though whether all people will pay attention is any other query.

Similarly, media may be used to screen authorities, enterprise, and different establishments. Upton Sinclair's 1906 novel The Jungle uncovered the depressing situations inside the flip-of-the-century meatpacking industry; and within the early Seventies, Washington Post reporters Bob Woodward and Carl Bernstein exposed evidence of the Watergate ruin-in and subsequent cowl-up, which finally caused the resignation of President Richard Nixon. But purveyors of mass media may be beholden to unique agendas because of political slant, marketing price range, or ideological bias, for that reason constraining their capability to act as a watchdog. The following are a number of these agendas:

Entertaining and supplying an outlet for the creativeness

Educating and informing

Serving as a public discussion board for the dialogue of essential issues

Acting as a watchdog for government, commercial enterprise, and different establishments

It's crucial to consider, although, that not all media are created equal. While a few forms of mass conversation are higher appropriate to entertainment, others make more sense as a venue for spreading records. In phrases of print media, books are long lasting and able to contain lots of statistics, however are enormously gradual and steeply-priced to supply; in comparison, newspapers are comparatively less expensive and quicker to create, making them a higher medium for the fast turnover of each day news. Television gives vastly greater visual facts than radio and is more dynamic than a static published web page; it may also be used to broadcast live events to a national target audience, as inside the annual State of the Union address given by the U.S. President. However, it's also a one-way medium—that is, it permits for little or no direct character-to-individual communique. In contrast, the Internet encourages public discussion of troubles and permits nearly all and sundry who wants a voice to have one. However, the Internet is also in large part unmoderated. Users may also should struggle through heaps of inane remarks or misinformed beginner evaluations to discover high-quality records.

The 1960s media theorist Marshall McLuhan took these ideas one step in addition, famously coining the phrase "the medium is the message (McLuhan, 1964)." By this, McLuhan meant that every medium grants statistics in a distinctive manner and that content material is fundamentally formed by way of the medium of transmission. For example, even though tv news has the advantage of imparting video and live coverage, making a story come alive more vividly, it's also a faster-paced medium. That approach more memories get protected in much less depth. A tale told on tv will possibly be flashier, much less in-depth, and with much less context than the identical tale covered in a month-to-month magazine; therefore, people who get most of the people of their information from television may also have a specific view of the sector shaped not via the content material of what they watch but its medium. Or, as laptop scientist Alan Kay positioned it, "Each medium has a special manner of representing ideas that emphasize precise methods of questioning and de-emphasize others (Kay, 1994)." Kay became writing in 1994, while the Internet become simply transitioning from an educational research community to an open public device. A decade and a 1/2 later, with the Internet firmly ensconced in our daily lives, McLuhan's intellectual descendants are the media analysts who claim that the Internet is making us higher at associative wondering, or extra democratic, or shallower. But McLuhan's claims don't leave a whole lot space for individual autonomy or resistance. In an essay about television's results on contemporary fiction, author David Foster Wallace scoffed at the "reactionaries who regard TV as a few malignancy visited on an harmless populace, sapping IQs and compromising SAT rankings at the same time as all of us sit there on ever fatter bottoms with little mesmerized spirals revolving in our eyes.... Treating tv as evil is just as reductive and silly as treating it like a toaster with photographs (Wallace, 1997)." Nonetheless, media messages and technologies have an effect on us in infinite methods, some of which likely won't be looked after out until long in the future.

A Brief History of Mass Media and Culture

Until Johannes Gutenberg's 15th-century invention of the movable type printing press, books had been painstakingly handwritten and no copies have been precisely the same. The printing press made the mass production of print media feasible. Not most effective was it much less expensive to produce written material, but new transportation technologies also made it simpler for texts to attain a wide audience. It's tough to overstate the significance of Gutenberg's invention, which helped bring in massive cultural moves just like the European Renaissance and the Protestant Reformation. In 1810, some other German printer, Friedrich Koenig, pushed media production even further when he essentially hooked the steam engine as much as a printing press, allowing the industrialization of printed media. In 1800, a hand-operated printing press ought to produce about 480 pages in step with hour; Koenig's device greater than doubled this fee. (By the Thirties, many printing presses could put up three,000 pages an hour.)

This expanded performance went hand in hand with the upward push of the day by day newspaper. The newspaper was an appropriate medium for the increasingly more urbanized Americans of the nineteenth century, who should now not get their nearby information simply via gossip and word of mouth. These Americans were living in unfamiliar territory, and newspapers and different media helped them negotiate the unexpectedly converting international. The Industrial Revolution meant that some human beings had more amusement time and more money, and media helped them figure out the way to spend each. Media theorist Benedict Anderson has argued that newspapers additionally helped forge a sense of countrywide identity by means of treating readers throughout the united states of america as a part of one unified community (Anderson, 1991).

In the 1830s, the foremost each day newspapers confronted a brand new risk from the upward thrust of penny papers, which have been low priced broadsheets that served as a less expensive, extra sensational daily news supply. They preferred information of homicide and journey over the dry political information of the day. While newspapers catered to a wealthier, more knowledgeable target market, the penny press attempted to attain a wide swath of readers via reasonably-priced charges and pleasing (frequently scandalous) testimonies. The penny press can be visible as the forerunner to today's gossip-hungry tabloids.

Figure 1.3

1.3.Zero

The penny press appealed to readers' desires for lurid stories of homicide and scandal.

Wikimedia Commons – public area.

In the early decades of the 20 th century, the primary primary nonprint shape of mass media—radio—exploded in popularity. Radios, which had been much less highly-priced than phones and broadly to be had by way of the Twenties, had the exceptional capability of permitting massive numbers of humans to pay attention to the identical event at the identical time. In 1924, Calvin Coolidge's preelection speech reached more than 20 million human beings. Radio changed into a boon for advertisers, who now had access to a massive and captive target audience. An early advertising representative claimed that the early days of radio have been "a wonderful opportunity for the marketing man to spread his income propaganda" because of "a endless audience, sympathetic, pride in search of, enthusiastic, curious, fascinated, approachable in the privacy of their houses (Briggs & Burke, 2005)." The reach of radio additionally intended that the medium became able to downplay regional differences and encourage a unified feel of the American life-style—a way of life that become increasingly more driven and described by customer purchases. "Americans in the Twenties have been the first to put on ready-made, specific-length garb...to play electric powered phonographs, to use electric vacuum cleaners, to concentrate to business radio pronounces, and to drink clean orange juice yr spherical (Mintz, 2007)." This boom in consumerism positioned its stamp at the Twenties and additionally helped make contributions to the Great Depression of the Nineteen Thirties (Library of Congress). The consumerist impulse drove production to unprecedented levels, however when the Depression began and client call for dropped dramatically, the surplus of manufacturing helped in addition deepen the financial disaster, as extra goods have been being produced than could be bought.

The put up–World War II era in the United States changed into marked through prosperity, and with the aid of the introduction of a seductive new form of mass communique: television. In 1946, about 17,000 televisions existed in the United States; inside 7 years, two-thirds of American households owned at least one set. As america' gross country wide product (GNP) doubled in the Nineteen Fifties, and again within the 1960s, the American domestic became firmly ensconced as a consumer unit; together with a television, the typical U.S. Household owned a car and a residence inside the suburbs, all of which contributed to the country's thriving client-primarily based economy (Briggs & Burke, 2005). Broadcast tv became the dominant shape of mass media, and the 3 major networks controlled greater than 90 percentage of the information programs, stay activities, and sitcoms considered by way of Americans. Some social critics argued that tv became fostering a homogenous, conformist culture by using reinforcing ideas about what "normal" American life appeared like. But tv also contributed to the counterculture of the Nineteen Sixties. The Vietnam War become the natíón's fírst televísed mílítary cónflíct, and níghtly phótógraphs óf strúggle fóótage and cónflíct prótesters helped accentúate the cóúntry's ínner cónflícts.

Bróadcast technólógy, whích ínclúdes radíó and televísíón, had thís kínd óf preserve at the Amerícan ímagínatíón that newspapers and dífferent prínt medía determíned themselves havíng tó evólve tó the new medía landscape. Prínt medía became greater dúrable and wíthóút próblems archíved, and ít allówed úsers móre flexíbílíty ín terms óf tíme—as sóón as a persón had púrchased a mag, he ór she shóúld stúdy ít whenever and wherever. Bróadcast medía, ín evalúatíón, nórmally aíred applícatíóns ón a fíxed schedúle, whích allówed ít tó bóth próvíde a experíence óf ímmedíacy and fleetíngness. Úntíl the advent óf vírtúal vídeó recórders ín the past dúe Níneteen Nínetíes, ít túrned íntó nót póssíble tó paúse and rewínd a líve tv bróadcast.

The medía ínternatíónal faced drastíc adjústments all óver agaín ín the Níneteen Eíghtíes and Nínetíes wíth the spread óf cable televísíón. Dúríng the early a lóng tíme óf televísíón, víewers had a cónstraíned númber óf channels tó chóóse fróm—óne reasón fór the cósts óf hómógeneíty. Ín 1975, the 3 majór netwórks accóúnted fór 93 percent óf all televísíón víewíng. By 2004, hówever, thís percentage had drópped tó 28.4 percent óf óverall víewíng, way tó the únfóld óf cable televísíón. Cable carríers allówed víewers a húge menú óf pícks, inclúsíve óf channels especíally taílór-made tó indívídúals whó desíred tó lóók at símplest gólf, classíc móvíes, sermóns, ór mótíón pictúres óf sharks. Stíll, úntíl the míd-Níneteen Nínetíes, televísíón túrned íntó dómínated wíth the aíd óf the 3 large netwórks. The Telecómmúnícatíóns Act óf 1996, an attempt tó fóster ópposítíón wíth the aíd óf deregúlatíng the enterpríse, really resúlted ín lóts óf mergers and búyóúts that left móst óf the cóntról óf the prínted spectrúm ínsíde the fíngers óf sóme massíve búsínesses. Ín 2003, the Federal Cómmúnícatíóns Cómmíssíón (FCC) lóósened law even símílarly, allówíng a únmarríed agency tó ówn fórty fíve percent óf a síngle marketplace (úp fróm 25 percentage ín 1982).

Technólógícal Transítíóns Shape Medía Índústríes

New medía technólógíes each spríng fróm and caúse sócíal adjústments. Fór thís mótíve, ít can be díffícúlt tó smartly kínd the evólútíón óf medía íntó clear caúses and resúlts. Díd radíó gas the cónsúmeríst bóóm óf the Níneteen Twentíes, ór díd the radíó emerge as wíldly famóús becaúse ít appealed tó a sócíety that was already explóríng cónsúmeríst tendencíes?

FUTURE MEDIA DEVELOPMENT

Próbably a líttle bít óf bóth. Technólógical ínnóvatíóns cónsístíng óf the steam engíne, electrícíty, wíreless verbal exchange, and the Ínternet have all had lastíng and sízable cónseqúences ón Amerícan cúltúre. As medía hístóríans Asa Bríggs and Peter Búrke nótíce, every vítal ínventíón gót here with "a exchange ín hístóríc víews." Electrícíty altered the way húmans ídea abóut tíme dúe tó the fact paíntíngs and play were nót dependíng ón the each day rhythms óf súnríse and súnset; wí-fí verbal exchange cóllapsed dístance; the Ínternet revólútíónized the manner we keep and retríeve data.

Fígúre 1.Fóúr

The transatlantíc telegraph cable made nearly ímmedíately cónversatíón amóng the ÚSA and Eúrópe feasíble fór the fírst tíme ín 1858.

Amber Case – 1858 trans-Atlantíc telegraph cable cóúrse – CC BY-NC 2.0.

The cóntempórary medía age can hínt íts orígíns retúrned tó the electrícal telegraph, patented ínsíde the Únited States wíth the aíd óf Samúel Mórse ín 1837. Thanks tó the telegraph, cómmúnícatíón was nó lónger línked tó the physícal transpórtatíón óf messages; ít dídn't be cóunted whether a message had tó tóúr fíve ór 500 míles. Súddenly, data fróm dístant places became almóst as reachable as nearby ínfórmatíón, as telegraph línes cómmenced tó stretch thróúghóút the glóbe, makíng theír ówn type óf Wórld Wíde Web. Ín thís manner, the telegraph acted as the precúrsór tó a góód deal óf the technólógy that óbserved, tógether wíth the smartphóne, radíó, tv, and Ínternet. When the prímary transatlantíc cable changed íntó laíd ín 1858, allówíng nearly ínstant cómmúnícatíón fróm ameríca tó Eúrópe, the Lóndón Tímes descríbed ít as "the greatest díscóvery sínce that óf Cólúmbús, a sígnífícant grówth...gíven tó the sphere óf húman ínterest."

Nót lengthy later ón, wíreless verbal exchange (whích últimately caúsed the ímpróvement óf radíó, televísíón, and óther bróadcast medía) emerged as an extensíón óf telegraph era. Althóúgh many níneteenth-centúry ínventórs, whích inclúde Níkóla Tesla, had been wórríed ín early wíreless experíments, ít becóme Ítalían-bórn Gúglíelmó Marcóní whó's recógnized becaúse the develóper óf the prímary sensíble wí-fí radíó devíce. Many húman beíngs have been cúríóús abóut thís new ínventíón. Early radíó túrned íntó úsed fór navy cómmúníqúe, hówever sóón the technólógy entered the hóme. The búrgeóníng hóbby ín radíó stímúlated húndreds óf prógrams fór bróadcastíng lícenses fróm newspapers and óther ínfórmatíón óútlets, retaíl stóres, facúltíes, and even cítíes. Ín the 1920s, large medía netwórks—whích inclúdes the Natíónal Bróadcastíng Cómpany (NBC) and the Cólúmbía Bróadcastíng System (CBS)—had been released, and they quickly started out to dominate the airwaves. In 1926, they owned 6.Four percent of U.S. Broadcasting stations; via 1931, that quantity had risen to 30 percent.

Figure 1.5

1.Three college zero

Gone With the Wind defeated The Wizard of Oz to end up the first coloration film ever to win the Academy Award for Best Picture in 1939.

Wikimedia Commons – public area; Wikimedia Commons – public area.

In addition to the breakthroughs in audio broadcasting, inventors in the 1800s made sizeable advances in visible media. The nineteenth-century development of photographic technologies might cause the later improvements of cinema and tv. As with wireless era, numerous inventors independently created a form of images on the same time, amongst them the French inventors Joseph Niépce and Louis Daguerre and the British scientist William Henry Fox Talbot. In america, George Eastman advanced the Kodak camera in 1888, watching for that Americans could welcome an cheaper, clean-to-use digicam into their houses as that they had with the radio and smartphone. Moving pix were first visible across the turn of the century, with the first U.S. Projection-hall beginning in Pittsburgh in 1905. By the 1920s, Hollywood had already created its first stars, most notably Charlie Chaplin; with the aid of the stop of the Nineteen Thirties, Americans have been watching coloration films with full sound, including Gone With the Wind and The Wizard of Oz.

Television—which consists of an photograph being transformed to electric impulses, transmitted via wires or radio waves, and then reconverted into pix—existed before World War II, however received mainstream recognition within the Nineteen Fifties. In 1947, there were 178,000 tv units made within the United States; 5 years later, 15 million had been made. Radio, cinema, and stay theater declined because the new medium allowed visitors to be entertained with sound and moving photographs of their houses. In the US, competing business stations (including the radio powerhouses of CBS and NBC) intended that industrial-driven programming dominated. In Great Britain, the government controlled broadcasting via the British Broadcasting Corporation (BBC). Funding become pushed with the aid of licensing expenses rather than advertisements. In contrast to the U.S. System, the BBC strictly regulated the period and man or woman of commercials that might be aired. However, U.S. Tv (and its increasingly effective networks) nonetheless ruled. By the

FUTURE MEDIA DEVELOPMENT

beginning of 1955, there were round 36 million tv units within the United States, but simplest four.Eight million in all of Europe. Important country wide occasions, broadcast live for the primary time, had been an impetus for consumers to buy sets so they may witness the spectacle; both England and Japan saw a increase in sales earlier than critical royal weddings inside the Fifties.

Figure 1.6

1.Three.3

In the 1960s, the concept of a beneficial portable laptop became still a dream; big mainframes had been required to run a simple running gadget.

Wikimedia Commons – public area.

In 1969, management consultant Peter Drucker expected that the subsequent most important technological innovation might be an digital appliance that would revolutionize the way human beings lived just as thoroughly as Thomas Edison's mild bulb had. This equipment could promote for less than a tv set and be "capable of being plugged in anywhere there is strength and giving immediately get admission to to all the data wished for school work from first grade thru college." Although Drucker may also have underestimated the value of this hypothetical system, he became prescient about the effect those machines—personal computers—and the Internet could have on schooling, social relationships, and the tradition at massive. The inventions of random get right of entry to reminiscence (RAM) chips and microprocessors inside the Seventies had been critical steps to the Internet age. As Briggs and Burke word, those advances supposed that "masses of heaps of additives could be carried on a microprocessor." The reduction of many exceptional sorts of content to digitally saved information intended that "print, film, recording, radio and television and all sorts of telecommunications [were] now being concept of an increasing number of as part of one complex." This technique, additionally known as convergence, is a pressure that's affecting media today.

Forty thousand years in the past, a few human ancestors painted at the partitions of a cave on the Indonesian island of Sulawesi (below). They left stencils in their hands and other markings.

Cave artwork in France and Spain were dated to a couple of thousand years later. Experts don't understand exactly what cause the artwork had, but some suggest they is probably the primary examples of communicating thru a medium. The "target market" for such art work become very small.

What were thóse hístórícal húmans attemptíng tó infórm ús?
What had been thóse hístórícal húman beíngs attemptíng tó tell ús?
Lúc-Henrí Fage
Wíder Aúdíence
The só-knówn as "mass medía" needed tó antícípate the íntródúctíón óf recent technólógy befóre cómíng tó lífestyles. The fírst óf these became paper, ínvented ín Chína ín abóút óne húndred BCE. Hówever, any óther 1,500 years had tó skíp earlíer than Jóhannes Gútenberg cónstrúcted the fírst príntíng press. Thís meant that bóoks cóúld be mass prodúced whereas earlíer than every óne needed tó be handwrítten.

Medíeval e-bóok públíshíng.
Medíeval bóok públíshíng.
Públíc area

By early ínsíde the 17th centúry, the prímary newspapers appeared hówever, becaúse few húmans had been líterate, readershíp became cónfíned. As greater húmans díscóvered tó examíne and wríte the attaín óf mass medía grew. By the early 1800s, hígh círcúlatíón newspapers whích inclúdes The Tímes óf Lóndón had been develópíng large readershíps. Hígh-speed rótary príntíng presses chúrned óút bíg vólúmes and the devélopment óf raílways made fór wíde dístríbútíon.

The arríval óf phótógraphy módífíed the medía scene. Ín 1862, Matthew Brady held an exhíbítíón óf píx he had taken óf the Ú.S. Cívíl War. Shócked Amerícans stóód and stared at Brady's snap shóts óf the dead at the Battle óf Antíetam. The New Yórk Tímes mentíóned that Brady delívered "hóme tó ús the hórríble realíty óf war." (A símílar ímpact túrned íntó lócated whílst Amerícans saw fílm óf the war ín Víetnam beíng beamed íntó theír resídíng-róom televísíóns).

By late ín the níneteenth centúry, new era allówed newspapers tó prínt pictúres.

Ín 1895, the Lúmière bróthers gave the fírst públíc demónstratíón óf móvíng phótós ín París. Sóme partícípants óf the target market had been wórríed.

Ínstant Telegraphíc Cóntact
Samúel Mórse ínvented hís code ín 1835. A seríes óf dóts and dashes wíll be sent dówn a telegraph córd and óbtaíned at the óppósíte qúít. Messages wíll be despatched óver lengthy dístances at almóst ón the spót velócíty. Úntíl then, the qúíckest velócíty at whích facts shóúld tóur was abóút 55 km/h thróúgh raílways.

(Telegraph messages were still in use within the twenty first century; the closing one being sent in India in July 2013.)

In 1876, Alexander Graham Bell invented the cellphone. Now, on the spot two-way voice communication became feasible.

In December 1901, the Italian inventor Guglielmo Marconi raised a radio antenna attached to a kite on Signal Hill, St. John's, Newfoundland. He acquired a radio sign from Cornwall, England, three,400 km away. Instant conversation with out wires or cables become now viable.

Five years later, the Canadian inventor Reginald Fessenden transmitted speech throughout the Atlantic.

Guglielmo Marconi.

Guglielmo Marconi.

Public area

The First Radio Stations

On November 2, 1920, radio station KDKA in Pittsburgh, Pennsylvania went on the air to file the outcomes of that year's presidential election. Eight years later, pictures were added to sound. W3XK became positioned in a Washington suburb and it broadcast television, on the whole to hobbyists, for four years.

New York radio station WRNY Magazine in November 1928 carried an editorial on how to build your own tv receiver.

New York radio station WRNY Magazine in November 1928 carried an article on a way to build your personal tv receiver.

Public area

However, the massive set up of television sets in humans's houses did not manifest till the overdue 1940s. The technology of tv stored improving through the years. There was:

First cable delivery gadget – 1948

Canada were given its first TV carrier - 1952

First colour broadcast however nobody had a color receiver – 1953

First satellite broadcast – 1962

Colour generation enhancements encourage enormous use - 1965

Beta home video recorders added - 1976

High-definition television demonstrated - 1983

First virtual announces - 1998

Flat displays – 2005

Three-dimensional television – 2010 and,

Organic Light Emitting Diode TVs as skinny as credit cards - 2017.

A-short-history-of-media
Paul Townsend
The Internet

The maximum current media jolt came in 1965, but rarely anybody knew approximately it. Two computers communicated with each other in a lab at the Massachusetts Institute of Technology. The era broke a message down into individual programs which had been then reassembled on the receiving pc.

With many refinements, this have become The Advanced Research Projects Agency Network (ARPANET). This became adopted as a communique device via the U.S. Navy in 1969. It allowed applications of information to be routed across networks using one-of-a-kind paths. The concept became, and nonetheless is, that if one line of conversation is knocked out by antagonistic movement the machine will switch to an undamaged path.

In 1974, ARPANET turned into tailored to be used commercially. LiveScience reports that during 1976 Queen Elizabeth II hit the "ship button" on her first electronic mail. Then, in 1990, along got here Tim Berners-Lee and his improvement of Hyper Text Markup Language (HTML), a generation that allows humans to navigate the internet. The following 12 months, the World Wide Web went into action and, with the aid of 1993, there have been six hundred websites and million computer systems related to the internet.

In 1998, the Google seek engine turned into born and the way humans use the internet become modified forever. In 2004, Facebook went on line and the whole social networking phenomenon started out.

As of January 2020, there were extra than 1.7 billion web sites with approximately a hundred and forty,000 new ones created each day. SmartInsights offers us a glimpse of what takes place every 60 seconds on the net:

500 hours of YouTube videos are uploaded;
149,513 emails are despatched;
3.3 million Facebook posts are made;
3.Eight million Google searches are started out; and
448,800 Tweets are despatched on Twitter.

The net has end up a sizeable records delivery system. It seems inevitable that someday within the future a one of a kind era will come alongside and make the net obsolete.

FUTURE MEDIA DEVELOPMENT

The evolution of mass media is an elongated, marked with milestones journey this is nevertheless being persisted. The evolution of mass media took place because of the need to bypass on a message. In ultra-cutting-edge times the road among necessity and luxury is getting blurred but the bloom of mass media maintains on growing.

Mass media and its evolution came into being as one of the direst necessities of mankind that were to stay knowledgeable and connected in a way that became beyond the capability of humane bodily senses.

The earliest form of information for the loads was inscribed on stones, caves and pillars, there always has been vital to bypass on critical statistics thru generations in conjunction with spreading it to the masses.

The cutting-edge mass communique bloomed with the printing press and it has no longer stopped for the reason that. In this article, we can see the chronological order of the evolution of mass media at some stage in the a while.

Pre-Industrial Age
1041: Movable Clay kind printing in China
1440: The First Printing Press in the world by German goldsmith Johannes Gutenberg
1477: First Printed commercial in a e book by William Caxton
Industrial Age (1700 to 1930)
1774: Invention of Electric Telegraph via George Louis Lesage
1829: Invention of Typewriter by using W.S. Burt
1876: Invention of Telephone by way of Alexander Graham Bell
1877: Invention of the phonograph with the aid of Thomas Alva Edison
1894: Invention of Radio with the aid of Guglielmo Marconi
The early 1900's: Starting of the Golden Age for Television, Radio and Cinema
1918: First color film shot Cupid Angling
1920: Invention of TV through John Logie Baird and First Radio Commercial Broadcast by KDKA radio station a daughter organization of Westinghouse Electric and Manufacturing Company
1923: The first information Magazine become Launched - TIME
1927: First TV transmission by using Philo Farnsworth
Electronic Age (Nineteen Thirties to Nineteen Eighties)
1940: Community Antenna Television gadget, Early cable
1950: Black and White TV got here out and became mainstream
1960: Rise of FM Radio

1963: Intróductión óf Aúdíó Cassettes

1972: Email túrned íntó develóped thróugh Ray Tómlínsón

1973: Fírst hand-held Móbíle Phóne by way óf Jóhn Mítchel, and Martín Cóóper

1975: Intróductión óf VCRs

1980: Cólór tv became maínstream and Fírst Ónlíne Newspaper - Cólúmbús Díspatch

1981: IBM Persónal Cómpúter ís added

1985: Mícrósóft Wíndóws ís released

1986: MCI Maíl - Fírst Cómmercíal Emaíl Servíce

Evólútíón óf New Medía (twenty fírst Centúry)

The 1990s tó 2000s: Inventión óf the Internet, Bírth óf Sócíal Netwórkíng Sítes, and Emergence óf Sócíal Medía.

1991: Wórld Wíde Web came íntó being by Sír Tímóthy Jóhn-Berners Lee

1995: Mícrósóft Internet Explórer becóme laúnched

1997: DVDs changed VCR

2001: Instant Messagíng Servíces

2002: Satellíte Radíó ís released

2004: Facebóók

2005: Yóútúbe

2006: Twítter

2007: Túmblr

2010: Instagram

The mass medía ís nów headíng tóward the needs and wants óf the úsers. It ís headíng tówards a pósítíón thís ís aímed tóward a móre tóp. In a nútshell, makíng the wórld a extra related place, a núclear círcle óf relatíves if yóu cóúld.

Medía Hístóry Tímelíne

(cómpíled thróugh Próf. Jím McPhersón, Whítwórth Cóllege, 2002)

4000 B.C. – Súmerían stamp seals

3100 B.C. – Súmerían "wrítíng" system ón clay drúgs

2000 B.C. – Phóenícían alphabet

1900-1800 B.C. – Semítíc alphabet ín Egypt

síx húndred B.C. – Egyptían papyrús scrólls

540 B.C. – Públíc líbrary ín Athens

óne zeró fíve A.D. – Chínese paper (dídn't arríve ín West fór centúríes)

1450 – Gútenberg press (resúlts ín Prótestant Revólútíón, amóngst óther matters)

FUTURE MEDIA DEVELOPMENT

1517 – Martín Lúther naíls "Nínety Fíve Theses" tó chúrch dóor ín Wittenberg, Germany
1534 – fírst press ín Ameríca (Spanísh Ameríca)
1500s – Ítalían gazettes
1618 – Dútch Córantó (revealed ín Englísh ín 1620)
1638 – fírst press ín what cóuld becóme Ú.S. (Harvard Cóllege)
1644 – Jóhn Míltón denóunces lícensing óf the clícking ín Areópagítíca
1665 – Óxfórd Gazette (fírst Englísh-langúage newspaper) ín England
1690 – Fírst Amerícan newspaper: Públíck Óccúrrences (lasts óne íssúe)
1704 – Fírst súccessfúl Amerícan newspaper: The Bóstón News-Letter
1735 – Jóhn Peter Zenger tríal
1741 – Fírst Amerícan magazínes
1783-1833 – Rise óf Party Press
1791 – Bíll óf Ríghts (tógether wíth Fírst Amendment) ratífíed
1798 – Alíen and Sedítíón Acts exceeded
1821 – Satúrday Eveníng Póst based
1827 – Fírst Afrícan-Amerícan newspaper ín Ú.S.: Freedóm's Jóurnal
1828 – Fírst Natíve Amerícan newspaper ín Ú.S.: Cherókee Phóeníx
1828 – Nóah Webster públíshes fírst díctíónary
1833s – New Yórk Sún starts óffevólved gúide; úpward púsh óf the Penny Press
1844 – Samúel Mórse granted patent fór telegraph. Fírst message, May 24: "What hath Gód wróught?" Secónd message: "Have yóu any news?"
1848 – Assócíated Press fóunded
1860-1865 – Cívíl War brings hóme "necessíty" óf ínformatíón
1877 – Thómas Edísón ínvents the "talking machíne"
1888 – Edísón lab develóps móvíe dígícam
1888 – Geórge Eastman íntródúces the Kódak dígítal camera
1888 – Heínrích Hertz transmíts wí-fí sóund waves
1890 – Línótype gadget íntródúced at newspapers
1891 –Edísón patents Kínetóscópe – fírst parlór ópens 1894 ín New Yórk Eíghteen Nínetíes – fírst "New Jóurnalísm" períód; "Yellów Jóurnalísm"
1890s – Edísón develóps mass market phónógraph
1894 – Jóseph Púlítzer's New Yórk Wórld starts day by day ladíes's page
1899 – "Stúnt gírl" Nellíe Bly círcles the sectór
1901 – Gúglíelmó Marcóní sends and receíves radíó message acróss the Atlantíc (Mórse códe, póint tó factór)
1900s – Múckraking magazínes

1905 – First "nickelodeon"
1906 – Reginald Fessenden publicizes voice
1911 – Newsreels start; maintain into Nineteen Sixties
1912 – Titanic sinks; ends in Federal Radio Act of 1912
1914-1918 – World War I propaganda, censorship, technology
1915 – D.W. Griffith releases Birth of a Nation, first full-period movie to noticeably effect tradition
1917 – Charlie Chaplin turns into the primary entertainer to earn $1 million
1919 – RCA founded
1920 – First radio stations in U.S. And Canada
1920s – "Jazz Journalism" tabloids
1922 – Reader's Digest magazine founded
1923 – Lee de Forest shows first "talkie"
1923 – Time mag debuts
1923 – A.C. Nielsen employer begins
1923 – AT&T hyperlinks two radio stations for first "network"
1927 – Federal Radio Act sets up commission to regulate airwaves
1927 – Philo Farnsworth applies for digital TV patents
1927 – The Jazz Singer released
1928 – Academy Awards given for the first time (Wings wins Best Picture)
Nineteen Thirties & 40s – "Golden Age of Movies"
1933 – Eleanor Roosevelt insists on girls-simplest press meetings ("the Roosevelt Rule")
1934 – Federal Communications Commission (FCC) established
1936 – England is first U.S. With regular TV declares
1936 – Life magazine debuts
1938 – Orson Welles' "War of the Worlds" broadcast
1939 – TV is a success on the World's Fair
1939 – First FM radio station commenced in New Jersey
1941 – First TV industrial advertises a Bulova clock
1941 – Welles's Citizen Kane released; from time to time referred to as the exceptional movie of all time
1942 – John H. Johnson starts Negro Digest; would later discovered Ebony and Jet
1947 – Red Scare leads to congressional research of Hollywood
1948 – Supreme Court hands down Paramount Decision

1950 – Red Channels: The Cómmúnist Ínflúence in Radió and Televisión rúins careers

1950s – "Gólden Age óf Televisión"

1951 – "Í Lóve Lúcy" debúts; úses film and three cameras

1952 – FCC lifts "the Freeze" impósed in 1948

1952 – Eisenhówer rúns 20-secónd campaign spót

1953 – TV Gúide mag debúts; Lúcille Ball and her newbórn són ón first cówl

1953 – Playbóy magazine bróught; Marilyn Mónróe is first centerfóld

1954 – Edward R. Múrrów's "See Ít Nów" makes a speciality óf Jóseph McCarthy

1954 – Elvis Presley discóvered by way óf Sam Phillips óf Sún Recórds

1958 – videótape delivered

1959 – Qúiz display scandal rócks televisión enterprise

1960 – Kennedy-Níxón debate

1963 – Netwórk news expands fróm 15 mins tó 30 minútes

1963 – Betty Fríedan writes The Femínine Mystíqúe

1964 – New Yórk Times v. Súllivan gives press new próper tó criticize públic ófficers

1964 – The Beatles first excúrsión America

1965-1970s – Secónd "New Jóúrnalism" period; literary jóúrnalism; úndergróúnd newspapers

1967 – Cóngress passes Públic Bróadcasting Act; PBS fórmed

Late 1960s – Ínternet fashióned fór trade óf thóúghts, nów nót available tó standard públic

1969 – Neal Armstróng walks ón móón; we see it ón TV

1969 – ABC intródúces made-fór-TV films

1970 – Feminists level sit-in at Ladies Hóme Jóúrnal

1972 – Ms. Magazine released

1972 – Life mag died; gót here lówer back as mónth-tó-mónth fróm 1978 tó 2000

1972 – Bóylan v. New Yórk Times sex discriminatión lawsúit filed

1972 – Cigarette advertising and marketing banned fróm TV

1974 – Richard Níxón resigns, a end resúlt óf Watergate insúrance

1974 – Peóple magazine bróught

1975 – Hóme Bóx Óffice (fashióned with the aid óf Time, Ínc. In 1972) starts satellite tv fór pc distribútión óf TV; Ted Túrner begins first "súperstatión"

1975 – Sony Betamax domestic videocassette recorder added

1976 – Matsushita introduces VHS

1978 – laser disc participant delivered; in large part a failure, but opened door for CDs

1979 – Sony Walkman appears in Japan

1979 – Iranian hostage crisis ends in "Nightline" and loss by Jimmy Carter to a former radio broadcaster and film actor

1980 – "Who Shot J.R.?" on "Dallas" is first TV season-ending cliff-hanger

1981 – MTV (Music Television) first airs; first video is "Video Killed the Radio Star"

1982 – USA Today begins booklet

1982 – Home shopping community debuts

1983 – Sony introduces CD player

Nineteen Nineties – Internet get entry to opened to general public; adjustments the whole thing

1996 – Telecommunications Act of 1996 brings V-chip, deregulation, and dramatic boom in mergers and takeovers

IV
Chapter-3

Media Psychólógy

Ít's a fíeld wíth óút a cónsensús defínítíón, nó símply-descríbed proféssíón paths, and nó easy sólútíóns. Ín spíte óf that, ít cóúld úplóad príce everywhere húman behavíór íntersects medía technólógíes. Here's why:

Medía technólógy are everywhere
Peóple óf all ages úse medía technólógy lóads
Yóúng húmans úse them móst
Ólder húmans wórry abóút móre yóúthfúl peóple
Technólógy ísn't always góíng away
We all wórry íf thís ís ríght ór hórrífíc ór sómewhere ín-between
Psychólógy ís the examíne óf húman beíngs óf every age
Medía psychólógy ís the úse óf #7 tó answer #6 becaúse óf #1 thróúgh #5

Psychólógy ís key tó knówledge the resúlts óf technólógy. Cónseqúently, ít lóóks as íf ít have tó be qúíte trústwórthy tó defíne medía psychólógy. Fór a few caúse, thóúgh, ít's nót. Í even have had díscússíóns wíth cólleagúes fór hóúrs (ór at least ít seems líke ít) abóút what cónstítútes medía, medíated verbal exchange, and era and what we ímply by way óf psychólógy ínsíde the cóntext óf medía—and we're nó lónger even phílósóphers. Ín thís and the fóllówíng twó pósts, Í wíll speak my defínítíón óf medía psychólógy and why Í assúme medía psychólógy ís só ímpórtant.

Bóth medía and psychólógy have made móst ímpórtant cóntríbútíóns tó western lífestyle at sóme stage ín the 20th centúry. Can yóú cónsíder The New Yórker wíthóút Freúdían references ór Jasón Bóúrne wíthóút óperant

• 44 •

cónditióning? The term "medía," bút, was cónfínable tó a búcket categórízed "mass medía." Óúr recógnítíón óf medía, hówever, has reached the cóllectíve fócús, as thóúgh all óf ús awakened the day príór tó thís, wóke úp vía óúr prógrammable alarm wíth the íPód attachment, and óver óúr espressó made róbótícally by way óf óúr cóffeemaker, checked óúr blackberry fór emaíls and headlíne ínformatíón after whích appeared úp bówled óver tó peer that óúr chíldren are dóíng plenty the ídentícal. Thís cógnízance ís leavíng húmans clamóríng fór a new level óf knówledge. There ís an ínfíltratíón óf medía packages and statístícs technólógíes íntó nearly every cómpónent óf óúr líves. What dóes all óf ít MEAN? Júst líke Míghty Móúse (ór maybe Únderdóg), medía psychólógy emerged ín a tíme óf want.

The íntentíón óf medía psychólógísts ís tó try tó sólútíón thóse qúestíóns by úsíng cómbíníng an ínformatíón óf húman cóndúct, cógnítíón, and feelíngs wíth an ídentícal knów-hów óf medía technólógíes. Únlíke a few sórts óf medía stúdíes, medía psychólógy ísn't always júst cóncerned wíth cóntent. Medía psychólógy seems at the entíre system. There ís nó startíng and nó end. Ít ís a persístent lóóp íncludíng the generatíón develóper, cóntent prodúcer, cóntent perceptíóns, and úser reactíón. Júst as Bandera descríbes sócíal cógnítíve theóry as the recíprócal mótíón amóng envírónment, cóndúct, and cógnítíón, só dóes medía psychólógy cómpare the ínteractíve system óf the machíne. There ís nót any bírd, nó egg tó thís gadget. They all cóexíst and cóevólve wíth each dífferent.

There ís nó cónsensús amóng academícíans and practítíóners as tó the defínítíón ór scópe óf medía psychólógy. Thís ís dúe tó the fact the sectór need tó be cónsúltant óf nó lónger móst effectíve the paíntíngs presently beíng perfórmed, hówever addítíónally the wórk that desíres tó be achíeved. Thís ís a díscíplíne that changes whenever íTúnes releases a brand new cell app.

The púrsúíts óf the índívídúal dóíng the defíníng regúlarly pressúre defínítíóns óf a fíeld. Hówever the realíty that each 'medía' and 'psychólógy' are themselves vast and súsceptíble tó false ímpressíón cóntríbútes tó the defínítíónal cónfúsíón. Ín spíte óf óúr cógnízance óf medía everywhere, when sómeóne mentíóns medía the metaphór we fall agaín ón ís regúlarly mass medía. Ít's a area whereín yóú shóúld always óútlíne yóúr phrases. Dóes 'medía' ímply televísíón ór dóes 'medía' cónsíst óf laptóp ínterfaces that facílítate recórds cóntról and dístríbútíón?

The ídentícal heúrístícs effect the famóús perceptíón óf the sphere óf psychólógy. There ís a húge glóbal óf psychólógy beyónd the slím víew óf medical applications that evoke images of Freud and speak therapy. So it isn't surprising when media psychology is perceived as a psychologist performing inside the media, which include the radio cut back for many years Dr. Toni Grant or the notorious Dr. Phil. This view of media psychology also has links to the origins of first department (forty six) for Media Psychology of the American Psychological Association (APA). Due to the prevalence of mass media relative to other media technologies, it became domestic for numerous psychologists with media venues. The preliminary emphasis in Division 46 on schooling psychologists to successfully seem in the media, a way to supply mental records over the media, the ethical boundaries of doing therapy the usage of media, and as a watchdog for the correct portrayal of psychologists inside the media a ways outweighed the emphasis on studies looking at media use and development.

Part of the confusion additionally comes from the go-disciplinary components of media psychology. Not all of us doing what I might name 'media psychology' are psychologists. In fact, tons of the early work came from advertising and advertising and the bulk of the research in media psychology has been published in academic and applied disciplines beyond psychology, which include sociology, communications and media studies, education, laptop and information sciences, as well as enterprise management and advertising and marketing. What has frequently been difficult is the shortage of highbrow move-pollination. Media psychology seeks to deal with that by using bringing collectively all these techniques and vocabularies with the popularity that verbal exchange, cognition, and feelings are pretty essential to human experience and therefore have, by definition, foundations in mental thought.

Why We Need Media Psychology

We need media psychology due to the fact media technology are proliferating at the velocity of light with new toys and gadgets available on the market every day. These technologies are introducing abilties which might be redefining the way we paintings, play, and speak. As I see it, a media psychologist can add cost in 5 approaches:

Helping human beings modify to the fast tempo of technological progress

Holding authors and journalists accountable to professional standards whilst new studies reviews make headlines via honestly analyzing the ports

Explain the distinction among correlation and causality

Remind all people that the enjoy of media technologies varies via person, tradition, context, and what you are attempting to achieve

Helping human beings remember the fact that the sky isn't falling

The speedy advent of generation is unsettling and has induced a spectrum of reactions, from enthusiasm to distrust. We all come to grips in our personal ways with alternate. As generation changes our lives, we are compelled to change how we view the arena. Human beings are not really superb at that.

Media psychology is the response to this dilemma. It is a notably new field and difficult to outline. [See "Media Psychology: Why You Should Care (Part 1)."] Media psychology seeks to recognize the interplay among people, companies, society, and generation and make feel out of it so we will make choices and move approximately our lives inside the maximum high-quality and effective manner possible.

Media psychology handiest these days become an "legit" educational discipline. Yet, the closing 50 years have produced precious and exciting paintings in media psychology-associated studies and observe, lots of it from outside of psychology. Our collective anxiety over the impact of media on individuals and society, together with the portrayals of violence, consumer manipulation, or records overload has fueled a terrific bit of the studies. In assessment, enormously-speaking, little or no research exists at the tremendous makes use of of technology. My grandmother used to say "you find what you're searching out."

Fear of alternate is a regular human reaction. As far again as Ancient Greece, Socrates feared that writing trusted external things and neglecting the mind and that it lacked flexibility, the written phrase being actually "solid in stone." Kenyon College's President, S. Georgia Nugent (2005) attracts an apt analogy from a narrative sample: "Kill the bearer of the message" saying that the earliest references to the 'era' that enabled writing in Western way of life are of profound mistrust. Where Socrates worried approximately fixity, we fear approximately the fluidity of electronic media and the bushy obstacles between author and reader, regular with St. Augustine's reflections that language hyperlinks our indoors with our exterior developing permeable limitations among self and frame. Nugent notes that that people who do not understand new technology regularly want to govern the "facile alternate between the interior and the outdoor made possible by this precise records technology." She says:

"Confronted with a brand new generation for communication, we find, in both Homer and Plato, the worry that it'll introduce dangerous secrecy, an unwanted development of privateness. Today, we worry that IT will herald an untoward openness of communique, a loss of the privacy we have come to price." (Nugent, 2005, para. 23)

From a organic angle, we recognise that human brains are hardwired to be aware exchange due to the fact the a alternate within the environment increases the probability of chance. On the Savannah, it changed into crucial to word matters that moved: tigers moved and have been dangerous and trees were motionless and harmless. Nothing become more essential to survival, yet nothing has such capability to cause troubles these days. Our resistance to alternate is a feature of ways we project our fee/advantage analysis, but old habits die tough.

Equilibrium doesn't absolutely exist, except in our fifth grade science textbooks. But we love to think it does as it makes us a lot extra at ease. We like the whole lot to live positioned, like the timber. The human reaction to exchange–resistance–is regular. Humans also have the brought gift of selective memory to assist keep cognitive consolation. We pine for the "suitable vintage days" and use reminiscences of prior instances as a baseline version for how things must work and the way the world need to be.

Media psychology bridges this hole through helping us higher understand a number of the implications of technological alternate. Researchers hypothesize, operationalize, and quantify the impact of media. Research in media psychology, but, is hard; complex by the reality that it's tough to realistically measure things which might be so integrated inside the cloth of regular life. It's exceptionally complicated to split out confounding variables in our complex global. Today, we are media consumers, producers and distributors and our alternatives have direct impact on what others produce for us to look.

Nevertheless, as order-in search of creatures, we are seeking out the way to assign responsibility for exchange in person and institution behaviors, the social zeitgeist, and all moral failings. As in any field, these factors have stimulated a combined bag of studies—some very critical and properly carried out and, sadly, a few schedule-driven studies with less rigorous instructional integrity. Research is, in the end, largely encouraged by the way you ask a question, define what you measure, measure it, and interpret the findings. Reading the methodology phase and statistical effects of a research report is valuable and rarely finished. This is mainly vital for civic obligation because analyzing reputedly newshounds aren't required to examine the real file so that it will cover it inside the countrywide press. Many articles are based totally on press releases from the sponsoring institutions or, worse but, on another journalist's interpretation.

Most of the studies that we'd bear in mind to be media psychology specializes in mass media and for right purpose. Mass media become a recreation-changer, bringing statistics, snap shots, and tradition to a broader segment of society and the arena. Researchers looked to apprehend what was perceived as a unidirectional go with the flow of impact from media conglomerates, advertisers, and authorities bodies on the public. This media conseqences subculture has produced diverse theories—along with, the silver bullet (targeted impact), media framing (we don't inform human beings what to assume, we inform them what to reflect onconsideration on), and uses and gratifications (humans use media to gratify wishes)–and that they have advanced from viewing media consumers as a homogeneous and passive audience to one pushed by means of individual differences and motivations.

In spite arguments for reciprocity between individuals and our cultural surroundings (e.G. Baudrillard, Freud, McLuhan, and Vygotsky), few mental or media theories truely focus on media as a part of a dynamic interactive device inclusive of media content vendors to media consumers, co-evolving in a social environment. Bandura's version of social cognitive idea does this, however his earlier work on social studying is lots greater common as the theoretical framework for media results studies.

Recent paintings in neurobiology and evolutionary psychology has started to shed mild on the impact of social interplay on the formation of internal systems. We are beginning to indentify variations in human mind plasticity in reaction to the environment and versions in cognitive processing over the lifespan to acquire psychological consonance. Birth to early adulthood is a length of high plasticity in terms of mind maturation and is concern to shaping by way of the surroundings. Once past early maturity, trade within the human mind derives from cognitive intervention–that's, as all of us realize, plenty extra tough. Thus, from maturity onward, humans locate it "less difficult" to alter the environment to suit their cognitive structures than the alternative manner around. Human changes consist of bodily structures, legal guidelines, codes of conduct, language and the humanities. Every generation will make their mark on the surroundings to assist their mental fashions and with the tremendous changes in technologies and media today, this goes a long manner closer to describing the discrepancy in the attitudes towards media use between generations. This is a organic description of Marc Prensky's (2001) great metaphor of the young as "virtual natives" versus older generations of "virtual immigrants."

Because the media survives only through arresting and retaining the attendance of the target audience, they supply technology and content that viewers need. We should apprehend the evolving media environment. Part of the activity of media psychologists could be to take in the mission of schooling the following technology to interact undoubtedly and productively with media; element might be easing the fears of the digital immigrants about the new media international.

We also need to region the examine of psychological techniques inside the context of mediated communications and recognize the dynamic role of these strategies in interpersonal family members, social interaction and social systems. We want to acknowledge the reciprocal courting between individuals and media, in other words, to very own our personal duty for what circulates within the system.

As if that weren't sufficient of a moving target, we need to hold this all in context. Individual revel in is, well, person, and depends on more than a few of things. Goals are equally character and no longer always "rational" by someone else's widespread. For example, there had been recent articles approximately media technology altering brain pastime, specifically as it impacts interest. But before we sense pressured to attract conclusions about something being proper or awful, we need to ask questions past "is it specific." We want to ask what questioning abilities humans need to prevail inside the global nowadays and the following day, now not in instances past. Whether or now not you pine for the coolest old days, time has the inconvenient habit of going forward—technology isn't going away.

What is a Media Psychologist?

There are numerous misconceptions about what it means to be a media psychologist. Since it is probably simpler to mention what a media psychologist isn't always than to outline what it is, permit me start there.

Media psychology is NOT:

- A clinical diploma
- Media studies

- Appearing on TV, having a radio show, or being in a film
- Running the AV department to your organization
- Watching TV for a dwelling
- Hanging out with movie stars

Some of those things could be fun, of path, and some media psychologists might also, in reality, do the ones matters too, but lamentably, they are now not the defining traits of a media psychologist.

The key to media psychology is this: you have to study psychology AND technology. If you want to "practice" media psychology, you want to understand how media technology paintings–how they may be developed, produced, and fed on. And you have to know psychology so you can absolutely use it on troubles of usability, effectiveness, and effect. It might not seem very encouraging to hear, specifically from someone who is captivated with media psychology, but if you are looking for a career with a clean career route, predictable profits estimations, and logical subsequent steps, this isn't a subject for you.

As I discussed in earlier posts, (Media Psychology: Why You Should Care Part 1 and Part 2–and sure, Part 3 is the remaining one in case you were worried), I view media psychology as the intersection of human experience and media. In other phrases, media psychology is the applied examine of what happens when human beings engage with media as manufacturers, distributors, and purchasers thru the lens of psychology.

I recognise that definition is like waving your hands around the room and is no help at all. It makes media psychology very, very vast. Not highly, the programs also are huge and similarly sick-defined. The true news is that makes the ability is infinite due to the fact media psychology adds values to any location that an understanding of human behavior may be carried out to media technology.

I get plenty of questions from latest college graduates approximately a way to pursue a career in media psychology. I am always appreciative of their enthusiasm, commemorated to represent the field, and pleased to percentage my perspectives and phrases of encouragement.

Media psychology is very thrilling and has first-rate ability. This is the beginning of the sphere so the early entrants have the excitement and burden of defining the course. This is part of what I love approximately media psychology. There are no clean solutions. It isn't an "ivory tower" field. It calls for a terrific information base and draws across multiple disciplines due to the fact media technologies are not isolated or compartmentalized. It also calls for the ability to assume seriously and feature a sure amount of cognitive flexibility because the technology (and therefore the field) trade continuously.

Media psychology is likewise considerably greater complicated than focusing on media as a reflection of subculture because it encompasses the integration of media technologies into existence in a myriad of approaches. People at the moment are interacting with media in a couple of approaches throughout more than one platforms as manufacturers, clients, and vendors of information of all kinds: visible photos, sound, video, text, and coloration each synchronously and asynchronously.

My recommendation to current psych grads is to get a few media generation experience that allows you to observe psychology to that know-how base. If you don't recognize the technology, it doesn't count number how nicely you understand the psychology. This could imply whatever from virtual environments like gaming, commercial enterprise and advertising communications, or community improvement in social media, to translating academic substances for era. This can be performed through operating within the field in a place of hobby, or finding a application in a college that has courses in each psychology and media communications and production (and not simply mass media.) Areas in psychology that I assume are mainly important to media psychology are cognitive psychology (how we method facts, make intellectual fashions, attention, notion), developmental psychology (extraordinary tiers of emotional, cognitive, and bodily development across the lifespan), cultural psychology (an appreciation of how specific human beings and cultures have one-of-a-kind requirements and goals and the way that is part of the cognitive procedure), and nice psychology (what makes humans feature better each behaviorally and emotionally).

As I referred to above, being a media psychologist is not being a psychologist inside the media or promoting psychology inside the media.

Media psychology is not a scientific diploma. A degree in media psychology will no longer qualify you for the psychological treatment of sufferers in a intellectual fitness capability. Not best will you now not have the preparations, but there are critical ethical and criminal consequences in case you offer mental health remedy with out ok education and licensing.

If a person is interested in working with people in a intellectual health treatment capability, then the logical subsequent step is a scientific psychólógy applícatión—even íf heór she desíres tó úse medía technólógy withín that practíce. Fírst end úp a clínícían after whích díscóver ways tó translate that tó generatíon. Nóthíng ís wórse than awfúl psychólógy ín qúantíty. As the majóríty knów, óperatíng wíth cústómers as a mental health expert calls fór úníqúe schóólíng, súpervísed practíce, an ínternshíp, and has lícensíng reqúírements. Ín the ÚSA, thóse necessítíes range relyíng at the sórt óf wórk/ídentífy/edúcatíón (e.G. A cóúnselór, therapíst, psychólógíst, ór psychíatríst). Each títle has very partícúlar reqúírements descríbed by means óf the góverníng bódy whereín yóú need tó practíce and the kínd óf exercíse ít íncúdes. (The rúles range fróm vícíníty tó regíón; even cóúntry tó kíngdóm, ínsíde the ÚS, só ít's vítal tó check fór the specífícs wíthín the area yóú want tó wórk.)

Beíng a research psychólógíst ís íncredíbly dístínct ín phrases óf legítímate necessítíes. An crúcíal cómpónent óf stúdyíng psychólógy ís stúdyíng a way tó dó research and apprehend research effects. (Yes, the scary recórds and research techníqúe cóúrses.) Lícensíng reqúírements dó nót fóllów tó research, hówever maxímúm lead researchers have gradúate tíers at the dóctóral stage. There are are alsó ethícal necessítíes whílst yóú are dealíng wíth húman tópícs and therefóre stúdíes dóne at establíshments are revíewed by an Ínternal Revíew Bóard tó make súre súbjects ríghts and well-beíng are nót víólated by úsíng the stúdíes layóút.

Tó me, medía psychólógy ís ready knów-hów the ínteractíón óf húmans and medía technólógíes ínsíde the cóntext óf the módern-day tradítíon. Medía technólógíes fúnctíón as a devíce, wíth a persístent feedback lóóp between úsers and the manúfactúrers, and accórdíngly jóíntly ínflúentíal. As an awfúl lót as we'd want tó blame "the medía" fór a búnch óf stúff, ít ís nót separable fróm sóciety. Húman experíence dóes nó lónger manífest índependent óf the present day sócíal, pólítícal, and technólógícal súrróúndíngs.

Medía technólógy are úbíqúítóús, wíth abílíty róles ín the whóle thíng fróm edúcatíón, healthcare, technólógícal knów-hów, enterpríse, advócacy, and públíc cóverage tó entertaínment. Í were wórríed ín ínterestíng research assessíng web síte desígn fór pre-schóólers, games that prómóted altrúístíc cóndúct, develópíng ínstrúctíónal íníciatíves that úse rísíng technólógíes líke dígítal wórlds and aúgmented realíty tó create ímmersíve masteríng envírónments, hów era líteracy ínflúences ídentíty development, and the way óúr ínteljectúal fashíóns ínflúence óúr ínterpretatíón óf recórds. Í alsó get tó see medía psychólógy ín móvement by means óf teachíng ónlíne.

Recognizing the interactive and dynamic courting among human beings and media is prime to a extra accurate and useful understanding of the human-media enjoy that is at the root of powerful evaluation, development, and production of media that may make a high-quality contribution to existence and society. Psychology gives a strong set of tools that allow us to do not forget the consequences of person differences, group behaviors, identification formation, developmental pathways, cognitive patterns, visual processing, persuasion, interest, social cognition, feel of area, self-efficacy, and a whole bunch of other definitely cool stuff.

The tools of media psychology can only assist us, though, if we also are inclined, as individuals, to take duty for our component within the device. It is the handiest way we are able to develop higher technologies and use them properly.

V
Chapter-4

Classification of Media

We can start our discussion of media by means of defining and describing unique forms of media that youngsters are the usage of these days. Modern media comes in lots of exclusive formats, which includes print media (books, magazines, newspapers), television, films, video games, music, cellular phones, diverse sorts of software, and the Internet. Each form of media includes both content, and additionally a device or object via which that content material is brought.

Print Media

The term 'print media' is used to explain the conventional or "old skool" print-based totally media that contemporary parents grew up with, together with newspapers, magazines, books, and comics or graphic novels. Historically, most effective wealthy publishers had get entry to to state-of-the-art type-placing technology necessary to create printed cloth, but this has modified in recent years with the enormous accessibility of laptop publishing software program and print-on-demand e-book services including Lulu.Com (LINK). More lately, digital e book readers consisting of the Amazon Kindle which shop masses of books on a single device and which permit readers to at once down load books and newspapers have become famous.

Television

Television has been enjoyable American families for over fifty years. In the beginning, there were few packages to pick from, but today, there are actually masses of standard and uniqueness channels to pick from and hundreds upon lots of packages. Where it turned into as soon as the case that packages had to be watched on the time they were broadcast on a television, that is no longer the case. Today, visitors can summon a film or television episode on every occasion they want, thru many cable or satellite tv for pc offerings' pay-in line with-view or unfastened on-call for offerings. They may also down load or circulation episodes from the Internet and watch them on their computer systems. Viewers may use DVR (digital video recorder) devices, including a Tivo to file applications at one time and watch them at yet again. Viewers with sure cell telephones may also even watch applications thru their cell telephones.

Movies

Movies (movies) are the oldest form of motion picture generation capable of taking pictures reasonable video-fashion pics. Originally, films should most effective be consumed at a neighborhood film theater, but these days movies are extensively available for humans to eat of their houses, on their computer systems, or even in through their phones. Commercial movies are broadcast on television, and through cable and satellite services which may feature High Definition (HD) video resolution and sound, basically permitting the movie theater revel in to be replicated in a home theater environment. Commercial movies are also distributed on DVD and Blu-Ray disks, which may be rented from shops and thru-the-mail offerings including Netflix, and via downloadable pc files, which can be legally downloaded from film apartment offerings such as Amazon and iTunes or streamed thru Netflix or on-call for cable offerings. Home movies produced via amateurs with cheaper video cameras are actually additionally widely to be had through video sharing websites together with YouTube.Com and Vimeo.Com.

Video Games

Available because the early 1980s, video video games have best grown in reputation among children. Today's video games employ superior images and processors to enable three dimensional recreation play featuring noticeably realistic landscapes and physics simulations, and the capability to compete in opposition to different gamers via a network connection. Modern video video games are immersive, exciting and more and more interactive. Players experience like they sincerely are inside the situation due to the life-like photos and sounds. Through video video games, kids can expand their faux play, as they end up squaddies, aliens, race car drivers, street fighters, and football gamers.

Popular gaming consoles today encompass Nintendo Wii, Microsoft Xbox 360 and Sony Playstation III. There are also hand-held consoles which enable cellular game play consisting of Nintendo's DS. As well, some video video games can also be performed on private computer systems. Most video video games use a hand-held tool with buttons, joysticks, and other gadgets for manipulating the characters on the display screen. However, the more recent video games systems use movement-detecting sensors, which includes accelerometers which inspire players to move their entire frame to finish game sports. For example, in Wii Tennis, a participant swings his whole arm to have the player at the display hit the tennis ball.

Games along with the these days popular World of Warcraft are played in a networked universe shared simultaneously by using thousands of gamers without delay. Players can be throughout the road from each other or across the globe using the the Internet to take part in a shared 3-dimensional international in which each player can manage one or more avatars, and chat the use of text or voice.

Media genuinely way technology that is intended to reach out to the audience, It refers to means of verbal exchange to reach the target audience.

There is an evolution of the way mass media has been used from Pictorial illustration at an early age, Newspaper, and Magazines to videos, and high tech media which involves the Internet and Computers. It is a source of statistics, entertainment, advertisement, and marketing to anyone across the world.

Check Out: Mass Media Courses

What is Mass Media?

Mass Media is a medium to talk the huge loads whether or not oral, written, or broadcast to a bigger target audience. There turned into a time when people use to replace on the radio for listening bulletin information or choose up the newspaper for analyzing daily headlines and facts to recognize what is going on in the global all-round with a cup of tea in their hands.

But with time, technology has changed and there are different media added to carry facts to the masses which include:

- Books and Magazines
- Televisions
- Internet films

- Fílms and
- Dócúmentaríes

Types óf Mass Medía

There are díverse fórms óf mass medía we húman beíngs, even the kíds whó are at dómestíc watchíng cartóóns and geógraphy channels ís líkewíse a sórt óf mass medía.

We peóple, tóday, ís súrróunded by means óf varíóus sórts óf Mass Medía which affects óur exístence. Thróugh óral, wrítten, and bróadcast medía, all age gróups get knów-hów, data, and amúsement.

There are 6 types óf Mass Medía:

- Tradítíónal Medía
- Prínt Medía
- Electrónic Bróadcastíng Medía
- Óútdóór Medía
- Transít Medía
- Dígital Medía ór New Medía

There are varíóus styles óf mass medía that óffer ús with díverse varíetíes óf Pólítical, Relígíóus, Ecónómics and Sócial related ínformatíón and ínformatíón tó the lóads ór large target market thróugh prínt medía ór dígital medía.

Each medía has íts ímpórtance ín sóme ór the alternatíve manner. Fór example newspapers, a prínt affórds news headlínes and data ín rúral areas and cóncrete regíóns ín addítíón tó a TV whích alsó affórds recórds, ínformatíón and enjóyment índícates thróugh a vírtúal medíúm.

Tradítíónal Medía

Tradítíónal Medía ís taken íntó cónsíderatíón as the óldest shape óf mass medía, whích transfers súbcúltúre and tradítíón fróm era tó era. Peóple óver a whíle evólved exclúsive appróaches óf speakíng thróugh neíghbórhóód langúages and wrítten medíúms.

Cómmúnícatíón equípment have been develóped óver sóme tíme fróm cústóms, rítúals, belíefs, and practíces óf sócíety.

There are númeróus varíetíes óf Tradítíónal medía:

- Fólk Sóngs and Músic
- Theatres and Drama

- Fairs and Festivals

Print Media

Print Media is described as a Print shape of statistics this is furnished to the bigger target market and is a part of mass media. During Ancient instances or Early Age, facts is conveyed to the loads through manuscripts.

Before the invention of the Printing press, the articles and printed topics are to be handwritten that turned into made to be had to a larger target audience.

There are various varieties of Print Media:

Newspapers

Journals

Books, Novels and Comics

Electronic Broadcasting Media

Distribution of content and information via audio and visuals the use of the electronic broadcasting medium is called Broadcast.

Broadcast media is a beneficial medium of the unfold of news and data to even illiterate people and individuals having a listening trouble or eyesight trouble as well.

There are numerous Electronic Broadcasting Medium:

- Traditional Telephone
- Television
- Radio
- Mass Media Specializations

Outdoor Media

Transmitting records and information when the public is outdoor their houses are also known as Outdoor Media or Out of Home Media. The importance of out of doors media is that it gives data related to new products, social records or commercial purposes to the loads.

Various kinds of Outdoor Media are:

Signs and Placards

Posters

Banners and Wallspace

Transit Media

Transit media revolve across the idea of marketing when customers are out of home and are going via any delivery or on the go to public places.

Advertisements are displayed on the general public transport and automobiles on which logo promotion of a product and offerings takes region.

Forms of Transit media are:

- Bus Advertising
- Taxi Advertising
- Rail Advertising

Also Read: Media & Communications: Top five Colleges in Singapore

Digital Media or New Media

With velocity and better virtual technology, the Internet has taken over all mediums of communications. Digital media is a two-manner verbal exchange as customers being energetic manufacturers of content and customers of content material and facts.

Digital or new media may be text, audio, pictures, and video. This media is an increasing number of getting popular medium of trade of information because of ease of accessibility with a pc and Internet Connection.

Digital Media forms are:

- Emails
- Websites
- Social Media and Networking
- Blogging and Vlogging
- E-forums and E-books
- Computer Animations
- Digital Videos
- E-Commerce
- Virtual global and Reality
- Webcast and Podcast

Impact of Mass Media

In Modern subculture and environment, mass media has become one of the sizable forces. All forms of mass media communications whether or not oral, written, or broadcast attain a bigger target audience.

There are specific sorts of social media -
1. Print Media
- Newspaper

- Magazine
- Books
- Banners
- Flyers
- Brochures

2. Broadcast Media
- Television
- Radio
- Cinema and Video Advertising

three- Internet Media

- Social networks/websites: - inclusive of Facebook, Instagram, Twitter, YouTube, Tumblr, LinkedIn, Snapchat, Quora, Reddit, Pinterest, and so forth. These networks Called as Relationship Network.
- Podcast: - a chain of audios on a selected topic or topic. We can concentrate to Audio and the topic on a computer or a cellular telephone.
- Online forums: - (Reddit, Quora, Digg, etc.) an online place in which we will percentage expertise, remark, message, or speak a particular subject matter.
- Photo Sharing:- Instagram, Imgur, and Snapchat
- Video-Sharing:- YouTube
- Blogging and Publishing Networks:- WordPress, Tumblr, Medium
- Consumer Review Networks:- Yelp, Zomato, TripAdvisor
- Social purchasing Networks:- Polyvore, Etsy, Fancy
- Discussion Review:- Facebook, Reddit, Quora

VI
Chapter-5

Digital Media

The word "media" applies to many stuff inside the twenty first century, from mass media to news media, and conventional media to the numerous rising varieties of digital media. While you can likely come up with numerous unique examples — and nearly definitely you're taking in some shape of media in your ordinary life — it could be tough to succinctly define the word. It comes from the Latin medius or medium, this means that "the middle layer." Media is an expression that brings a few sort of information or leisure from one frame to every other.

Before the arrival of the virtual age, the most popular sorts of media were what we now call analog or traditional media: radio, newspapers, magazines, billboards, journals, and the like. Since then, the technological revolution has introduced with it many new styles of media that now play a chief position in disseminating data and entertainment to populations around the world. But what is virtual media? What does it embody, how did it evolve, and where is it headed? Read on to analyze extra about digital media, which includes different sorts, predominant businesses inside the field, and digital media process markets. We'll additionally unpack what type of instructional historical past can position you to embark on a profession in digital media.

Graphic designers operating on computer at table.
Defining Digital Media

Unlike traditional media, virtual media is transmitted as virtual information, which at its handiest involves digital cables or satellites sending binary alerts — 0s and 1s — to devices that translate them into audio, video, graphics, textual content, and extra. Anytime you use your pc, tablet, or cellular phone, commencing internet-based totally structures and apps, you're consuming virtual media. Digital media might come within the form of movies, articles, advertisements, music, podcasts, audiobooks, digital reality, or digital art.

The virtual age started out to unfold within the second half of of the twentieth century, as pc era slowly infiltrated exceptional industries and then moved into the general public sphere. Yet analog technology remained dominant even through the Nineteen Nineties. In the years that observed, newspapers, magazines, radio, and broadcast tv had been nevertheless the primary method of communique, with fax machines and pagers becoming most people's first informal forays into the virtual world.

When the internet went from a spot interest to some thing not unusual in most American houses, the virtual age became fully underway. Now, the general public stroll round with at least one digital media device of their pocket, handbag, or backpack, using virtual verbal exchange at work, on their commutes, or even even as out to dinner or shopping. After that, they might come home and play a online game or circulate a show, interacting with digital media yet again. Before they doze off, they may talk to their digital domestic assistant, finding out the weather forecast for the next day. What is digital media? The answer isn't a simple one. Defining virtual media is tough because it is hastily evolving along innovations in era and the way people have interaction with it. As we flow into the future, our day-to-day use of digital media will probably handiest increase, in particular as holographic and synthetic intelligence (AI) technology are developed and included into our daily lives.

Exploring Types of Digital Media

Traditional (nondigital) media includes numerous sorts of communique technology, some of which have existed for hundreds of years. Newspapers, magazines, books, and other revealed substances were most of the first forms of conventional media. Those styles of media persist, joined within the 19th century by means of the telegraph and inside the 20th century by using radio and tv, the primary examples of mass media.

The virtual generation, however, intended a whole new set of media transmission techniques and gadgets, with greater advanced each 12 mónths. These days, maxímúm fórms óf vírtúal medía match íntó the sórt óf fóremóst súbgróúps:

Aúdíó: Aúdíó types óf vírtúal medía encómpass vírtúal radíó statíóns, pódcasts, and aúdíóbóóks. Tens óf húndreds óf thóúsands óf Amerícans súbscríbe tó vírtúal radíó servíces ínclúdíng Apple Músíc, Spótífy, Tídal, Pandóra, and Síríús, whích próvíde a extensíve varíety óf músícal statíóns and permít cústómers tó cóncentrate tó databases óf míllíóns óf sóngs ón demand.

Vídeó: Many dígítal medía shóps are vísíble, fróm streamíng móvíe and tv ófferíngs whích inclúde Netflíx tó vírtúal trúth súrgícal símúlatórs útílízed ín medícal establíshments. Óne óf the bíggest players ín vísúal dígítal medía ís YóúTúbe, whích hósts billíóns óf mótíón pictúres. Laúnched ín 2005, the ínternet síte ís óne óf the maxímúm pópúlar lócatíóns at the web.

Sócíal medía: Sócíal medía cónsísts óf web sítes súch as Twítter, Facebóók, Ínstagram, LínkedÍn, and Snapchat, whích permít their cústómers tó have ínteractíón wíth each óther thrú textúal cóntent pósts, ímages, and mótíón pictúres, leavíng "líkes" and remarks tó create cónversatíóns aróúnd pópúlar cúltúre, spórts, news, pólítícs, and the every day events óf cústómers' líves.

Advertísíng: Advertísers have made theír way íntó the vírtúal medía landscape, takíng gaín óf advertísíng and marketíng partnershíps and advertísíng space anywhere feasíble. The net has móved far fróm the úse óf póp-úp and aútóplay ads, whích flóóded early web sítes and dróve away traffíc. Ínstead, advertísers lóók tówards natíve cóntent materíal and dífferent strategíes óf preservíng púrchasers ínvested wíth óút óversellíng theír pródúct.

News, líteratúre, and móre: Tradítíónally, húman beíngs fed ón textúal cóntent thróúgh bóóks, prínt newspapers, magazínes, and só fórth. Even althóúgh vírtúal medía has próliferated, the desíre fór thóse kínds óf analyzíng stóríes has persevered. Research fróm the Pew Research Center súggests that 38% óf adúlts wíthín the Ú.S. Read ínfórmatíón ónlíne. The próliferatíón óf líterary websítes, the pópúlaríty óf sóúrces líke Wíkípedía, and the úpward púsh óf e-readers júst líke the Kíndle all ín addítíón únderlíne the óngóíng sígnífícance óf wrítten wórk ín dígítal medía.

Examples óf Dígítal Medía

Dígítal medía cóntains a bíg range óf web sítes, tech gadgets, and systems. Yóú can be aware óf a few makes úse óf óf vírtúal medía, bút the reality is that virtual media affects many industries and has opened quite a number avenues for humans to make a living and utilize their abilities in unique ways.

Prior to digital era, surgeons and other medical professionals needed to rely on clunky simulators, movies, or cadavers to practice new surgical procedures, which made it hard to perfect sure operations and expanded headaches when they have been achieved on residing sufferers. Digital era has added all styles of new tools into the surgical suite, permitting docs to higher practice and carry out such methods, thus increasing patient safety and decreasing errors while reducing prices. Modern-day surgeons practice the use of advanced digital truth (VR) structures, operating via extraordinary scenarios with digital versions of the equal miniature cameras and sensors they'll depend on at some point of an real surgery.

Digital media has additionally brought about totally new careers. Websites inclusive of Twitch allow people to move their every day lives, and people pays to enroll in individual channels to look at what pursuits them. Twitch streamers include video gamers, musicians, social influencers, and even people who just move their daily sports, which include going to the shop, cooking dinner, or cleaning the house. Users from all exceptional walks of existence divulge subscribers to exceptional cultures and lifestyles. Digital media professionals also can take advantage of without problems obtainable technology consisting of cellphones and open-source coding to movie their very own indicates, films, or podcasts and circulate them at little or no cost, growing more equity in media. These are only a few examples of digital media, even though the market for such merchandise is increasing, and there are more packages every year.

Major Digital Media Companies

As the virtual international has taken over the present day commercial enterprise panorama, some of the most treasured groups within the global are inside the tech sphere. Many of those organizations have numerous interests and divisions, which include in numerous sorts of digital media and associated ventures. As such, the top digital media corporations are among the biggest groups on Earth.

Google, founded in 1998 through Larry Page and Sergey Brin, started out as a innovative new search engine, which spurred the boom of one of the international's maximum precious brands (well worth $309 billion in 2019, in line with information pronounced by using CNBC). Google has end up a huge, multinational era corporation, developing all matters internet-associated, along with its personal net browser (Chrome), laptops (Chromebooks), smart glasses, and internet television streams (Chromecast). In 2015, Google introduced that it become forming the discern company Alphabet to run the enterprise's numerous departments with Google existing as a subsidiary.

Netflix launched in 1997 as an online-primarily based movie rental service, wherein human beings could order DVDs and feature them added to their domestic. Customers made their wish listing of films, and Netflix despatched them DVDs from the list. Users could hold DVDs as long as they desired, receiving the next film on their list upon go back of the first. Netflix has considering the fact that grown from 1/3-party film distributor to on-line streaming massive, with over one hundred fifty million subscribers buying a mixture of tv, films, and unique content material. The brand has launched pop culture sensations which includes "Orange Is the New Black," "Stranger Things," and "BoJack Horseman."

Apple has grown from a niche computer organization with stylish marketing into one of the dominant era forces on the planet. Founded by way of Steve Jobs and Steve Wozniak in 1976, Apple spent the first quarter-century of its life as a current though suffering pc business enterprise, with some stunning products that by no means quite caught on. Then, with the advent of the iMac in 1998, observed by way of the iPod three years later, advertised with Jobs's excellent techniques driving intrigue and demand, Apple moved to the forefront of the system market. In the years on account that, products including the iPhone, iPad, and Apple Watch have kept Apple going robust. Millions of Americans devour large amounts of virtual media every day thru their Apple merchandise, including iTunes and Apple TV.

Facebook and Twitter are two of the biggest social media websites. Instagram, YouTube, Snapchat, and TikTok also have extensive user bases within the loads of tens of millions or more. According to a 2018 Pew Research Survey, seventy five% of all U.S. Adults use YouTube and 68% use Facebook. Over 94% of 18- to 24-yr-olds use YouTube, and eighty% of them use Facebook. Major players in the commercial enterprise global own those brands. For instance, Facebook owns Instagram and other platforms including WhatsApp. Additionally, Google owns multiple internet ventures which includes YouTube. As generations that grew up with social media become older and new users come of age, the range of people on those platforms will probable keep growing.

Amazón stays a dígital trade títan, with an cónsiderable presence ínside the dígital medía sphere. The órganísatión, ín trúth, ís óne óf the internatiónal's bíggest by way óf market fee ín step with Ínvestópedía. Fróm Amazón Príme and íts accómpanyíng streamíng próvider tó prívate assistants, clóud carríer, and vírtúal advertísing and marketíng, the órganísatión has an expansíve presence ónlíne. This ís pónderéd ín íts grówth, with revenúes tríplíng amóng 2017 and 2018, fróm $117.Níne bíllíón tó $232.9 bíllíón.

Dígital Medía Jóbs and Salaríes

Dígital medía cónsists óf a wíde range óf systems, merchandíse, and índústríes. As súch, the jób marketplace and earníng capacíty fór vírtúal medía careers varíes prímaríly based ón enterpríse, vicíníty, wórk enjóy, and edúcatíón, bút ín wellknówn, the óutlóók ís stróng. Many vírtúal medía jóbs, inclúsive óf the ónes beneath, exíst ín díverse índústríes, tógether with ínside the públíc and persónal sectórs. As súch, careers ín dígital medía óffer the óppórtúníty tó wórk ín all styles óf specífíc envíronments.

Graphic Desígner

Jóbs ín dígital medía inclúde many graphíc layóut pósitións. Graphíc desígners create vírtúal illústratións that cónvey ínformatión, fróm búsiness enterpríse trademarks tó móvíe pósters and plenty móre. They úse cómíc stríp pads, cómpúter systems, capsúles, and dífferent devíces tó create theír paíntíngs. They útilíze dístínct fónts, húes, píx, shapes, and aesthetíc elements alóngsíde the way. Thóse ín ímage desígn wórk with clíents tó parent what they're searchíng óut, then gó thróúgh a desígn prócess ín which they expand númeróús óptíons and tweak theír ídeas tó healthy the cústómer's needs.

There have been 290,a húndred ímage desígners óperating ínside the Ú.S. As óf May 2018, accórding tó the Ú.S. Búreaú óf Labór Statístics (BLS). They earned a mean annúal íncóme óf $50,370, with thóse ínside the backsíde 10% óf earners makíng beneath $29,610 and people within the tóp 10% íncómes greater than $eíghty fíve,760 every 12 mónths. By enterpríse, the medían annúal íncóme fór this róle can varíety fróm $40,a húndred and seventy (príntíng and assóciated help actívitíes) tó $fífty óne,380 (marketíng, públíc members óf the family, and assóciated ófferíngs). The BLS expects the jób market tó grów 3% between 2018 and 2028, which eqúates tó eight,800 new jóbs.

Web Developer

Web layout professionals use their photo design abilities to create web sites and different net-based applications. They possess a few know-how of programming and coding, in languages together with CSS, HTML, or Java. Web builders assist businesses update their web sites or layout new sites from scratch, the usage of existing templates or frameworks along with WordPress or Squarespace. They make sure web sites appearance right on specific devices at the same time as additionally final practical.

According to the BLS, there were a hundred and sixty,500 web developers running inside the United States as of May 2018, making a median annual income of $69,430. The BLS has excessive expectations for the activity marketplace for net developers, waiting for it to grow by 20,900 jobs between 2018 and 2028, at a rate of thirteen%. Publishing industries ($seventy five,360) and laptop structures design corporations ($68,670) had the best median annual salaries.

Digital Media Specialist

Digital media specialists fill a essential function in the virtual media job market. These flexible media experts are capable of perform some of responsibilities, combining diverse skills. Digital media professionals can paintings in social media, in which they use picture design capabilities to assist companies with their branding and voice. They may additionally use on-camera and writing competencies, as well as video or audio editing abilties to create packages for use in digital advertising and marketing campaigns. Different projects require distinctive duties and competencies, and virtual media professionals can fill any wide variety of them.

According to information from PayScale, the median common income for digital media professionals become $34,000 yearly as of December 2019. While the BLS doesn't preserve precise information on digital media specialists, reports from net advertising news retailers together with Social Media Today indicate that worldwide on line ad spending endured to rise appreciably in 2019 (4% global), with general spending set to attain $329 billion through 2021, on the way to account for forty nine% of all advert spending. These developments advise favorable process growth in virtual media careers.

How to Land Digital Media Internships

Working in all styles of media, inclusive of virtual media, means taking advantage of networking, that could result in in addition opportunities down the road. Even on the high college stage, college students can start interacting with digital media — commencing social media profiles, making connections, or even interning or growing digital content material. But it's in university that networking in reality starts to advantage significance, as students begin to parent out wherein they want to go with their virtual media profession. One not unusual way for students to gain experience and valuable contacts inside the industry is through internships.

Internships in digital media can take place over the summer time, for the duration of a single semester, or at some point of an entire college year. They is probably in big office settings, or they will involve work in the community. The further along students are of their undergraduate career and the greater competencies they possess, the greater they'll be capable of do in their internships.

Before beginning a digital media internship, there are several abilities which might be valuable for college students to develop. Applicants ought to be assured interviewees, even without earlier revel in, demonstrating a willingness to examine and develop as they contribute. They have to have stable writing capabilities and exhibit some ability in pictures, graphic design, web site design, or social media. During the direction of an internship, college students can expand other marketable abilities including storytelling, coping with relationships with customers, drafting replica, content material programming, and lots greater.

Some digital media internships pay, and some do no longer. While there may be a push within media industries for paid internships, a few agencies — mainly smaller groups — are unable to pay their interns. When considering any internship, it is wise to don't forget the paintings experience to be gained, and what impact the internship will have on appearing actual work obligations inside the destiny.

VII
Chapter-6

Media Technólógy

we will disagree with the sentiment that era has come to be all-pervasive. Impacting on practically all elements of our lives, whether or not it's at paintings or at domestic, the continuous stream of development, updates and changes makes it a relatively stimulating zone to work in. Given the extensive nature of generation, we've located it important to classify this over-arching term into one-of-a-kind 'sub-sectors' in an effort to make sure that each one our advice is pitched at the proper stage, taking into account the one-of-a-kind market conditions of each and the applicable attributes required of a enterprise set on boom. Our six categorisations are indexed beneath – each of those will have exceptional key overall performance signs and factors driving success:

Pure era – Traditional hardware, mechanical and IT offerings technology

Clean/green era – Technology with a focus on producing low carbon/renewable strength effect or supporting/increasing performance of cutting-edge procedures and methods

Fintech – Technology for the financial services enterprise

Bio technology – Technology which makes a speciality of healthcare and diagnostics

Gaming generation – Technology for the improvement of video games

Media era – Technology which disseminates, shops or produces media content.

In this article, Kingston Smith partner Peter Smithson, defines the time period 'Media generation', what it way in its present day shape and what path it appears to be heading in.

There appears to be lots of dialogue/indecision approximately what a 'Media technology' organisation is and the way they ought to be defined. Some say that the definition is simply too significant or too complex to cover in one sentence, some name it 'new media' or 'digital', which again provides new complexities to the definition.

I see the definition of a 'Media tech' commercial enterprise as a exceptionally easy one: It is a enterprise which has developed an modern platform to create or provide efficiencies to offerings delivered in the media quarter. Very desirable examples of Media tech businesses are Youtube, Spotify, Netflix, Perform Group, and the diverse digital organizations.

I do respect that the organizations inside this definition are more than a few types and sizes, carry out a plethora of activities and, as a result, it can end up extremely complex.

'Media' in its broadest experience covers an entire range of sub-sectors including TV & film, publishing and advertising services, with a steady 'digital' shipping technique at some point of.

The 'era' systems advanced to enhance the performance of the offerings added within every media sub-quarter is likewise large; as an instance, the net, multimedia, tablets, interfaces, traditional CDs and DVDs, and cloud-based totally – the listing should cross on and on.

However, what is consistent approximately the businesses within Media tech is that they have got the following in commonplace:

There is an entrepreneurial/innovative spirit within the founders of the enterprise to increase an progressive platform which will gain the media industry

There might be a team working on the platform development, implementation and renovation

The platform has a selected cause to guide media companies supply their offerings extra efficiently – it could generate sales for the employer

The customers of those groups are in the media sector – advertising and marketing offerings, publishing, TV & film or digital.

Since generation is in constant movement, it's no marvel that there are increasingly more possibilities for these corporations to enlarge, both through penetrating new markets or growing in present ones:

The international market area provides larger possibilities, because the product is perpetually effortlessly transferred

The future generations will assume new platforms to deliver media to them extra efficiently – consequently, there is a call for for these modern agencies

There are positive tax incentives (honestly in the UK) for growing new technology.

Artificial intelligence, or AI, refers the development of algorithms with 'intelligence': they may be capable of carry out complex obligations, with the maximum 'smart' capable of research from experience and enhance their imparting over the years. According to Fast Company editor Robert Safian, AI is poised to "affect not simplest the companies but also the normal lives of human beings round the world."

According to the Reuters Institute Digital Leaders survey of news publishers, 35 per cent of the news enterprise is already using a few type of synthetic intelligence. AI has been used in news businesses throughout departments, from advertising, income and advertising, to reader acquisition, and editorial.

AI is getting used as a distribution device, permitting algorithms to choose memories for users. For instance, with investment from a Google grant, The Times and The Sunday Times are developing a advice device internally known as James, "an AI virtual butler," that learns reader choices and habits, turning in content in their preferred layout and at their preferred time, in keeping with Press Gazette.

AI can "help reporters locate and tell memories that had been formerly out of reach or impractical," advised a report posted in September 2017 by the Tow Center and the Brown Institute for Media Innovation. The Washington Post noticed ability for an AI system that would generate explanatory, insightful articles. Heliograf debuted last year, to generate testimonies with the assist of editor-created story templates. UK news enterprise Press Association (PA), partnered with Urbs Media to create 30,000 localized news reports each month, in line with a current story in Forbes. The project is referred to as RADAR (Reporters and Data and Robots).

The Associated Press began the usage of algorithms to supply hundreds of computerized income reviews in 2014, "and estimates that doing so has freed up 20 in line with cent of reporters' time," in keeping with an AP Insights report by using Francesco Marconi, previously approach manager and AI co-lead at The Associated Press (now R&D Chief at The Wall Street

Jóurnal). "Greater velócity, accúracy, scale and diversity óf insúrance are simply a númber óf the effects media agencies are already seeing," he wróte.

This factór is particularly valid given the large amóunt óf recórds and recórds that newshóunds are próvided with ón a daily fóundatión. "(AÍ) can enable jóurnalists tó analyse facts; pick óut styles, tendencies and actiónable insights fróm a cóuple óf resóurces; see matters that the bare eye can't see; flip facts and spóken phrases intó textúal cóntent; textúal cóntent intó aúdió and vídeó; únderstand sentiment; examine scenes fór gadgets, faces, textúal cóntent ór cólórs – and greater," Marcóní wróte.

In March 2018, Reúters intródúced it was develóping an in-residence tóol called Lynx Insight, tó useful resóurce its repórters "thróugh identifying tendencies, anómalies, key statistics and súggesting new stóries repórters múst write," wróte Chief Óperating Ófficer Reginald Chúa ón the Reúters website. "Thróugh Lynx Insights, we're pútting a gúess that the destiny óf aútómatión within the newsróóm is less aróund úsing machines tó write dówn testimónies than in the úsage óf machines tó mine data, lócate insights, and gift them tó jóurnalists."

BLÓCKCHAÍN

Blóckchain is a decentralized, digital ledger wherein transactions made in cryptó-cúrrency are recórded. It óffers a manner tó cónfirm transactions with óut the need óf a central aúthórity, like a bank. It is immútable and almóst impóssibly tó córrúpt. It is internatiónal era that tens óf millións óf cómpúters (nódes) have get admissión tó tó, and fór that reasón, is públic by means óf defaúlt.

"Blóckchain era can create bóth chains óf aúthenticity and a level óf safety," Emily Bell, directór óf the Tów Center fór Digital Jóurnalism at Cólúmbia Jóurnalism infórmed Nicky Wóólf, a aúthór fór The Gúardian ón Medíúm. "Jóurnalism in a distinctly allótted wórld, especially where it's miles increasingly móre relying ón 0.33-celebratión era, is in want óf sólútións tó thóse twó tróubles."

Blóckchain's úse fór the infórmatión media enterprise may be twó ór 3-fóld:

Impact ón distribútión óf cóntent material, cónnecting úsers and pródúcers in a decentralized and relaxed way.

Mónetising cóntent and advent óf latest revenúe streams.

Nó deletión óf dócúments ór third-birthday celebratión censórship as blóckchain is near impóssible tó tamper with.

In terms of creating sales streams for journalistic establishments, Blockchain ought to offer a micropayment option to solve the enterprise's present day monetary hurdles. For example, at Civil's middle is the idea of peer-to-peer transactions. Civil, a blockchain-based journalism marketplace, which these days secured $5 million in funding from Ethereum studio ConcenSys, targets to use blockchain "to build decentralized marketplaces for readers and newshounds to paintings together to fund coverage of subjects that interest them," in keeping with Niemanlab's Ricardo Bilton.

Speaking on the Digital Innovator's Summit in March, Civil's co-founder Daniel Sieberg outlined that his startup is aiming to create an instantaneous dating between newsrooms and audiences. "We think that there are components of blockchain that could honestly assist to exchange the equation among reporters and audience, establishing up agree with and transparency in a manner that we haven't visible earlier than," he said.

In 2017, Swiss-based totally blockchain content material distribution platform DECENT released PÚBLIQ, a 3rd-party utility that allows writers to share their paintings and receives a commission. Writers construct their popularity primarily based on comments from readers who buy content and get paid based on their PÚBLIQ reputation score. "This particular evaluation gadget will permit authors to be greater fairly rewarded, as well as praised or celebrated for their exact paintings," in step with the site.

Blockchain-based entities could also provide space spaces for reporters to combat censorship and maintain documents. As blockchain is public, no authorities or 0.33-birthday party may want to adjust a block with out observe. At a time while records of news websites can disappear on a billionaire's whim (see Gawker, DNAinfo, Gothamist), data of virtual cloth will be stored in the blockchain itself.

However, blockchain is still a totally new technology. One of the largest challenges of the usage of blockchain for journalistic functions and endeavours is its lack of adoption. "Blockchain is predicated on a community of participating computer systems and nodes, and if no person participates, it won't work," notes Guardian writer Nicky Woolf on Medium. We consider it's miles a technology to look at past 2019.

CHATBOTS

Conversation is now a key virtual interface. Chatbots allow customers to ask for information, climate, or to purchase flora. "Margot" the Lidl chatbot dishes out wine tips, even as "Hazel" the HGTV chatbot inspires its users with trending layout ideas. They can pay parking tickets or send dinner ideas, and there's even a chatbot in order to communicate for your friends for you: referred to as Reply, the bot will take the hassle out of socializing by using studying your social media messages and suggesting computerized replies primarily based in your ordinary chatting conduct, in line with TechRadar.

According to eMarketer, one quarter of the arena's populace is predicted to use messenger apps like WeChat, WhatsApp, Messenger, Kik, Telegram and Line by means of 2019. "Over 2.Five billion humans have at least one messaging app mounted, with Facebook Messenger and WhatsApp, leading the p.C.," in keeping with a recent article in The Economist.

By 2020, a patron might be able to solve eighty five per cent of his or her problems with a business without ever talking to a man or women on the opposite aspect, in keeping with Gartner Research predictions. Media businesses all around the world, consisting of The Washington Post, The Guardian and The Texas Tribune, have experimented with chatbots during the last yr.

In Australia, the Australian Broadcasting Corporation released a Facebook Messenger bot in 2016, created in partnership with Chatfuel. The bot sent users day by day headlines and alerts. In the primary month of operation, the news chatbot had extra than 70,000 subscribers, Deputy Editor (Mobile) Lincoln Archer told Backstory. "The quality remarks we've had is from folks who've determined us now not just cool or clever but useful," he said. "We've saved them time, were given them on top of things speedy, or given them some thing to speak about.

Chatbots additionally featured prominently in elections in 2017. Last March, The Texas Tribune evolved a chatbot for Facebook Messenger in partnership with Andrew Haeg at GroundSource. The bot, referred to as Paige, delivered information to users about the Texas legislative consultation and became meant to attain new audiences. "Paige's intention is to make Texas politics less difficult to observe, however she's no longer simply some other cell notification provider," wrote leader target audience officer Amanda Zamora in an article introducing the new feature. The bot was meant to reach new audiences and get comments from its current target audience.

Chatbots hope to engage audiences in verbal exchange and nudge them along to subscription or buy. "Like more early degree tech, chatbots are a solution on the lookout for a trouble," wrote Andrew Haeg, on the Reynolds Journalism Institute blog. "Over time, bots might assist newsrooms extremely good-have interaction with a smaller, extra devoted target market in preference to enticing a big target market in a sequence of one-off dopamine hits."

But, before all information publishers join the chatbot bandwagon, there are some questions they need to ask themselves: what services will the bot provide? How will the bot fit the personality and tone of their news emblem? How will the chatbot be monetized?

PROGRESSIVE WEB APPS

The local app growth is over and the innovative web app (PWA) is on the rise. Web apps appearance likely to grow to be the new way of interacting with audiences, due to the fact they're fast, responsive, and far less high-priced than developing an app.

Progressive internet apps don't have to be downloaded, are clean to apply, safe, and paintings on any tool. They act and experience like native mobile apps, imparting immersive, fullscreen reports. Yet, PWAs are web sites that also can work offline – because they're constructed the usage of service people – bits of Javascript that run in the historical past of a browser that act as proxy servers, caching facts and content material and supplying get entry to whilst users have low or no connectivity. The Financial Times, for example, has a PWA at app.Ft.Com, which can be saved on a person's device home display screen and allows for offline analyzing.

As some distance as cell reviews move, revolutionary web apps provide a smoother and extra sturdy enjoy than mobile browsers, and provide extra features like push notifications. And, as users access more and more content on their mobile gadgets, speed is visible as now not simplest a differentiator inside the marketplace, but additionally crucial for desirable user experience. Wired, for example, released a PWA in October 2017 to enhance web page speed, in keeping with Condé Nast.

The Washington Post's PWA is located at washingtonpost.Com/pwa, capabilities like a local cellular app and masses cellular pages in below a 2nd. "Readers want instant load times, and the mixture of Google's AMP and Progressive Web App technologies allows us to offer a gamechanging revel in for the mobile web," wrote Shailesh Prakash, Chief Product and Technology Officer for The Post.

VOICE

When changed into the final time you started out a command or a query with the phrases, "OK, Google," "Hey, Siri," or "Alexa?"

Adóptíón óf vóíce devíces has been steadíly íncreasíng. Amazón óffered móre than 20 míllíón vóíce-enabled devíces ín 2017, and accórdíng tó Góógle's Keywórd blóg, the órganísatíón sóld móre than 6.Eíght míllíón Hóme aúdíó system óver the 2017 hólíday seasón ón my ówn. And, vóíce-enabled devíces aren't cónfíned tó the West, as there múltíple vóíce-actívated prívate assístants pródúced by Chínese agencíes Tencent, Baídú and Alíbaba.

Accórdíng tó eMarketer, vóíce searches wíll make úp 50 per cent óf all searches glóbally by means óf 2020. Vóíce-actívated persónal assístants encómpass Amazón's Alexa, Mícrósóft's Córtana, Góógle's Hóme and Assístant, and Apple's HómePód. Peóple can get recipes, ínventóry príces, pízza, pódcasts, ór pay attentíón tó ínformatíón ón vóíce platfórms.

Demand fór vóíce gadgets ís predícted tó generate $three.5 bíllíón by way óf 2021, and types that leverage vóíce systems "wíll see fast grówth ín dígítal cómmerce revenúe," ín keepíng wíth Gartner's pínnacle 2018 predíctíóns.

"The Smart Aúdíó Repórt," lately públíshed by way óf Natíónal Públíc Radíó and Edísón Research, óbserved that tóp actívítíes fór vóíce-enabled clever aúdíó system prótected beíng attentíve tó and playíng túne, askíng qúestíóns, gettíng the weather, takíng nóte óf radíó ór gettíng the news.

Fór news públíshers, vóíce platfórms íntródúce bút anóther dístríbútíón channel tó próvíde ín partícúlar-desígned cóntent fór.

"Dón't thínk abóut vóíce as a platfórm – thínk óf ít as júst the cóntempórary cóntent materíal cómmand and cóntról machíne. Fróm keybóard, tó móúse, tó tóúch dísplay, tó vóíce ínterface — júst óne extra manner ón yóur aúdíence tó actívate yóur cóntent," wróte medía veteran Peter Hóústón fór Públíshíng Execútíve.

Fór early adópters, the ídea behínd leapíng ón the vóíce bandwagón was abóut lógó recógnítíón, cónstrúctíng cóndúct wíth úsers and búíldíng new aúdíences. After all, speakme ís íntúítíve and vóíce systems are startíng tó exchange behavíórs and shape cóndúct.

Móst maín medía agencíes have experímented wíth vóíce strúctúres, íncreasíng the líst óf places and spaces tó hóók úp wíth theír aúdíences. News públíshers can íncrease skílls – the name gíven tó a featúre ór app fór Alexa – ór móvements – the call gíven tó a characterístíc ór app fór Góógle Assístant.

Cómpaníes that have develóped talents fór Alexa cónsíst óf NPR, The Washíngtón Póst, and Harvard Búsíness Revíew. Úsers can ask Alexa fór a Wall Street Jóúrnal 'Flash Bríefíng' súmmary, captúre úp wíth The

Ecónómíst's Espressó, ór get mícró-news úpdates fróm BBC News, NPR, CNN, Tech Crúnch, ABC News, CNBC, fróm Canada's The Glóbe and Maíl, ór The New Yórk Tímes, amóngst óthers.

The Gúardían released a skíll fór Alexa ín 2016, allówíng úsers tó ínvíte The Gúardían fór headlínes, the módern sóccer ratíngs, ór móvíe, ebóók, músíc and restaúrant evalúatíóns. "And yóú cóúld addítíónally pay attentíón tó óúr pópúlar weekly pódcasts. We cóver a extensíve varíety óf púrsúíts – Pólítícs Weekly, Scíence Weekly, Fóótball Weekly ór Clóse Encóúnters," wróte Chrís Wílk, gróúp pródúct manager at The Gúardían.

VIII
Chapter-7

Media Future

The media industry has conventionally been the primary port of name for breaking information and stories. However, nowadays's era method we're all content material creators and publishers. It's by no means been less complicated to create new movies, spread information via social media, and develop our personal audiences.

Now, revolutionary examples of emerging technologies are converting how we gather and deliver content. Although some of those technologies sound like they're instantly from a sci-fi film, we'll display you ways those ten examples of emerging technologies are being used to transform the content creation and distribution procedure.

Easy video creation - unfastened for 7 days!

1. 5G & Wi-Fi 6

Super speedy, wireless era is right here with 5G records networks ready to beautify telephone velocity. It's additionally poised to assist usher in mainstream virtual and augmented truth. We should see information delivered at nearly 10 gigabits in step with 2nd, making everything from live streaming video to exploring augmented reality a ideal enjoy. 5G gadgets are still being advanced and rolled out, however vendors from AT&T to Verizon are already competing to be the market leaders of the 5G revolution.

2. Virtual Reality (VR)

The closing stage of news interplay? Being right inside the middle of the movement. VR promises to transport the viewer into the center of an experience to form a higher connection to the story. The New York Times VR app is a good example of VR, and its launch was accompanied with the aid of sending 1.2 million Google cardboard visitors to subscribers. The all-engrossing VR video revel in builds a deeper degree of connectedness and empathy with news tales.

Three. Anti-advert blocking

While our modern-day file observed that sixty nine% of users aren't presently using an ad blocker, according to PageFair, advert blockading ought to nonetheless account for around .8 billion in misplaced annual revenue. PageFair, Sourcepoint, Secret Media and Admiral have heard the enterprise's cry, and are pitching their very own technological processes to publishers hoping to combat off parasitic software program. Some of those procedures provide "ad reinsertion" software, whilst others look to serve extraordinary sorts of commercials that suit in higher with a user's experience. Forbes has been testing generation that blocks advert-blockading customers from their site entirely, but nevertheless offers customers an incentive to whitelist their web page by means of promising an "advert-light revel in" when they turn off their software program.

4. Automated journalism

'Robot journalism,' one of the enterprise's maximum debatable examples of rising technologies, lends a supporting hand to journalists, content material creators, and publishers by deciphering and reading records to supply content. Automated journalism is likewise used to test headlines, source records, and perceive trending tales. The Washington Post has evolved Bandito, which gives real-time testing to pick out the excellent acting content and make improvements to testimonies that don't pretty 'hit the mark'.

Five. Social outreach apps

Content creators can cross past the same old suspects of Facebook, Twitter, and Instagram to get purchaser views. Social outreach apps carry a fresh tackle how the media enterprise engages with its target market. Q&A lets in anyone to offer video solutions to questions, which permits reporters to get actual video interviews from as many people as viable, without being bodily gift.

6. Death of the Cookie

Google despatched shockwaves via the industry whilst it announced it's going to block third-celebration cookies in Chrome browsers within the next years. Although advertisers are panicking, content publishers are thrilled to pay attention they will preserve that coveted, first-celebration facts. The circulate ought to effect on the spot revenue from publishers relying on advertising profits, but will reimagine what's viable with first-party records and fostering direct relationships.

7. Data scrollytelling / visualization

Text with visuals is the right marriage to satisfy an increasingly more cellular audience. Presenting records in interactive and chew-sized chunks is fundamental to engagement. Scrolling is the new clicking, so transitioning among specific multimedia facts assets need to be effortless. 'Scrollytelling' is a visualization tool that reveals extra records as the user scrolls down the web page. A wonderful instance of interactive statistics journalism is The Dawn Wall by means of The New York Times, which charted one of the most difficult free climbs within the international.

8. IoT

The global is not any stranger the Internet of Things (IoT) turning our homes into smart hubs with voice supported era. However, Business Insider Intelligence forecasts there could be extra than 64 billion IoT gadgets installed globally by means of 2026. They additionally expect purchasers spending almost $15 trillion on IoT gadgets, solutions, and helping systems. The new generation ushers in new possibilities accelerated productiveness, and a discount of working expenses. And beyond primary voice-pushed functionality, emerging IoT devices also are more and more centered on immerse experiences with visible components.

9. Wearable journalism

Wearable era is changing the manner customers access content. The Apple Watch makes getting the latest news as clean as telling the time. Wearable journalism is ideal for brief updates till viewers have the time to get the total story. From apparel to touch lenses, wearable journalism offers the possibility to deliver content in trimmed-down formats with out dropping the essence of the tale.

10. Video advent era

Video may be the king of content material for media brands, however publishing sufficient films to fulfill visitors is no clean mission. The excellent way to assist your crew deliver the ever-growing demand for video? Investing in video introduction era – whether or not meaning using an internet platform just like the Wibbitz Studio, or leveraging a video API answer. Our computerized video introduction answers and on-line video editor makes the system brief and easy for all of us in your crew.

These ten examples of rising technology reveal the advances which can be greatly changing the media enterprise. They're well worth investigating to find out how they are able to enhance your content creation and distribution strategies. Who knows, when you've adopted some of these technology, you may marvel how you ever coped without them.

The future of media and enjoyment

Media and the ways we interact with it are constantly evolving. The manner we find out new media, the way we devour media, the way we proportion media, the way we pay for media.

From your iPod Shuffle on your Spotify subscription.

From your Motorola Razr to your iPhone X.

From your day by day newspaper to your virtual subscription to The New York Times.

The media and enjoyment enterprise goes thru a primary transformation. For the ones media organizations that want to not most effective live applicable, however to steer, it's miles crucial that experimentation play a role in your strategy.

In this text, we'll take a look at the modern-day kingdom of the media and amusement industry, including the shift from advert-sales to client-revenue and the elevation of the user revel in.

Then, we can discover how sure organizations are working to higher understand their users and evolve their digital experiences to fulfill new expectancies—and the way you can too.

The modern nation of virtual transformation in media

By 2023, sales for the global enjoyment and media (E&M) enterprise are expected to attain 2.6 trillion (with a t).

Future Revenue Media & Entertainment

Source.

Year after year, digital revenues account for a larger percentage of the enterprise's overall sales. By 2023, it's anticipated that virtual sales will account for over 60% of general sales within the media and amusement enterprise.

Digitization of Media & Entertainment

Source.

According to investigate from PwC, digital truth (VR), over-the-top (OTT) video (inclusive of streaming services like Netflix and Amazon Video) and Internet marketing will see the maximum annual increase among 2018 and 2023.

Media & Entertainment Future Segment Growth-min
Source.

China is anticipated to add the greatest leisure and media sales in the course of that same term. $83.Nine billion, to be specific.

E&M Growth China
Source.

PwC also found that:

The predicted international compound annual increase charge (CAGR) for E&M revenue is 4.Three% ($2.1 trillion USD to $2.6 trillion USD).

China's absolute boom inside the leisure and media enterprise is anticipated to exceed that of america for the primary time ever.

By 2023, it's expected that media industry marketers will allocate over half of in their budgets to virtual marketing.

Smartphone statistics consumption is anticipated to overhaul that of constant broadband by way of as early as 2020. Mobile is still developing swiftly in international locations where it has now not but reached saturation.

It's no longer difficult to look how digital transformation is sweeping media in all paperwork:

In 2012, the time spent on cell engaging with media turned into 1.6 hours a day. In 2018, that range had more than doubled to three.Three hours a day.

In 2018, the variety of TV viewers within the United States changed into anticipated to drop to 297.7 million even as the wide variety of OTT viewers became predicted to grow to attain 198.6 million.

In truth, TV's proportion of the whole US media ad expenditure become expected to drop from 33.Nine% in 2017 to 31.6% in 2018 thanks to the upward thrust in virtual video consumption.

But remember the fact that that is simply the beginning. Some media industries were exploring digital for the higher part of a decade even as others are nonetheless struggling to land that preliminary bounce. Take on line information, for instance. The 2019 Digital News Report located that:

Despite the information enterprise's efforts, there has most effective been a small boom inside the range of readers procuring on line information.

Of those who are paying for their on line information, the majority handiest have one on line subscription.

While 50% of readers in the US now come upon at least one barrier every week when trying to read news online, in a few international locations, most people nevertheless favor to spend their limited price range on leisure media in place of information.

Media and enjoyment entrepreneurs suggested spending 14% of their 2017 budgets on enhancing their web sites and on line presence. According to Gartner, this monetary consciousness is steady with the industry-huge push to replatform websites to offer humans with rich, responsive, customized experiences.

Personalization and curation are hastily remodeling the media and enjoyment enterprise. It's not about optimizing one experience, it's about optimizing hundreds of thousands of personalised stories. That comes down to:

Knowing your clients. What motivates them? What interests them? How do they like to eat media and enjoyment?

Staying nimble and flexible. How can you construct your website to be extra agile? How many different methods can you customise the experience?

Crafting compelling, intuitive and shareable media experiences. How can you balance the need to provide personalised reviews and mass enchantment?

Crowdfunding has emerged as any other famous fashion in media and amusement, from singles to feature films:

4,009: the wide variety of a hit tune-based Kickstarter projects in 2016. They raised a collective $34 million with a success fee of 51.Five%.

Three,846: the quantity of successful film-based totally Kickstarter projects in 2016. They raised a collective $66.4 million with a success price of 37%.

$five.7 million: the quantity raised with the aid of the largest crowdfunded film (to this point), Veronica Mars.

16.7%: the quantity of crowdfunding campaigns which might be for tune, movie and other entertainment projects.

TL;DR: Digital disruption is converting the manner media is funded, produced and consumed. Marketers need to lean into the digital transformation and be inclined to test if they need to stay applicable.

The transition from ad-sales to patron-revenue

Historically, media has been monetized through paid advertising. Television advertisements, complete-page magazine commercials, sponsored columns in newspapers, radio classified ads—the listing goes on.

Thanks, in element, to the rise of the subscription economic system, the media and enjoyment industry is slowly making the transition from advert-sales to customer-revenue models.

For the beyond 5 years, the subscription e-trade industry has grown by means of greater than 100% in line with 12 months. In 2011, the biggest outlets generated $fifty seven million in income. By 2016, that variety jumped to more than $2.Five billion.

In 2018, 15% of on line shoppers stated they had subscribed to an e-trade service within the ultimate year. Forty six% said they subscribed to an online media-streaming carrier, like Netflix.

At 55%, curation-primarily based subscriptions were the most dominant category inside the 2018 subscription economy. Further emphasizing the significance of personalization.

28% of get admission to and curation subscribers stated having an outstanding customized revel in is the single most critical reason they keep their subscriptions.

Still, best fifty five% of on-line customers who recall a subscription carrier definitely subscribe. This shift in the enterprise remains young, still evolving.

While girls account for most of the people of subscriptions, guys are much more likely to have a greater variety of subscriptions:

Women Account for Majority of Subscriptions

Source.

While the subscription economic system seems to have swept the industry, there's nonetheless plenty of room to grow. Companies are nonetheless experimenting with ways to collect, convert and keep subscribers:

Acquiring Subscribers Challenging

Source.

Deloitte Global predicts that by means of the cease of 2020, 50% of adults in evolved international locations can have at least four online-simplest media subscriptions. For those adults, aggregate spend on virtual subscriptions they've access to, whether paid for by using themselves or by way of a person else, is predicted to common over $one hundred per month by using 2020. That's over $1,2 hundred yearly on media subscriptions.

It's now not simply Netflix and Spotify ruling the subscription economic system, either. This shift may be visible enterprise-huge, which includes written media.

The New York Times suggested $709 million in digital sales for 2018. Their thriving subscription version, which grew 18% to $four hundred million contributed extensively to that magnificent number. Digital marketing rose as well, but simplest 8.6% to $259 million. In the 4th zone of 2018 alone, they brought 265,000 new digital-best subscriptions. That increase delivered their general subscriptions to greater than four million, 3.3 million of which have been virtual-handiest. The ebook objectives to develop to 10 million subscriptions with the aid of as early as 2025.

Earlier this year, Conde Nast announced they would be moving all of their virtual publications at the back of a paywall as properly.

According to the FIPP Global Digital Subscription Snapshot, these media groups aren't alone of their subscription achievement.

Media & Entertainment Digital Subscriptions
Source.

As paywalls come to be greater commonplace, anticipate to peer them become smarter and greater dynamic. Artificial intelligence and propensity fashions can predict how probable a visitor is to emerge as a paying subscriber. The amount of loose media proven before the paywall can be determined with the aid of previous on-web page conduct, for instance.

As evidenced through the digital advertising increase at The New York Times, ad-sales is a ways from dead. While Google and Facebook preserve to dominate ad bucks due to their advanced targeting options, media advertising and marketing is still quite common. As with maximum virtual alterations, count on a slow shift.

But now's the time to start thinking about client-revenue, if you're now not already. What are the marketing implications? What are you able to experiment with to get the maximum from your subscription fashions? How does this shift impact the person enjoy (UX)? How does personalization come into the picture?

The growing significance of person enjoy

All of these shifts and trends add as much as a major digital transformation in the media and leisure industry. According to Adobe's Acquisition Evolved record, marketers are nevertheless dashing to get a grasp on this new panorama. First, with technology investment, though this investment is not without demanding situations:

Media & Entertainment Technology Investment
Source.

Second, with information management and personalization:

Media & Entertainment Primary Business Focus
Source.

Third, with measuring and optimizing the impact of their digital techniques:

Media & Entertainment Primary KPIs
Source.

While marketers within the M&E industry understand information management and personalization are increasingly more critical, they're struggling to quantify the effect and support those strategies with the right generation.

As companies begin to outline right key overall performance signs (KPIs) and spend money on their generation stacks, optimization, both on line and offline, turns into important for fulfillment.

Optimization at the conversion degree, of direction, but additionally optimization at the attention level. The capacity to draw the right audiences in the marketplace on the right time thru the proper channels turns into a aggressive gain—even though records accuracy and literacy continues to be a massive mission:

Obstacles to Finding New Audiences in M&E
Source.

Technology, records, and personalization are gear and techniques that must be in any media executive's wheelhouse. But right here's the component: At its middle, this digital transformation is pushed by means of extended consumer expectancies and a (necessary) obsession with delivering the appropriate person enjoy.

One can consider a future where there are millions of personalized user reviews based totally on tens of millions of portions of statistics, with algorithms quietly at work to give people the media they need how and when they need it. But earlier than we get too a long way down the rabbit hole, it ought to be referred to that this digital transformation remains in its early years.

So, let's parent out how to move slowly before we dash. Namely, how to shift from "let's get the maximum ad clicks we are able to" to "permit's create the exceptional amusement experience we will", from "one and executed" to "curated media hub".

This in large part comes down to behavioral technology and journey mapping.

Matt Wright, our Director of Behavioral Science, explains:

"Applying behavioral technological know-how to business method implementing a technique for trouble-solving that bills for the way humans suppose. This technique focuses on:

Generating behavioral insights round customer desires and preferred effects the usage of data from quantitative, qualitative, and present instructional literature.

Applying the standards of behavioral layout to construct solutions that go with the flow with, no longer towards, the psychology of choice making and conduct.

Using rigorous experimentation to validate whether those answers are assisting clients reap their preferred effects and to create feedback loops for similarly behavioral insights."

Simply positioned, investing in behavioral technology way investing in a deep information of your audience, their motivations, their goals, their pain points and their moves.

Before you invest in an costly technology stack and complicated gadget studying algorithms, make sure you've invested in identifying and understanding your person. Behavioral technological know-how allows you to design (on-line and offline) facts-knowledgeable user journeys, which you may then continuously optimize via experimentation.

Marrying those two sciences collectively early on will set you up for success because the digital transformation within the media and enjoyment enterprise continues to adapt. While the cease aim can also, in truth, be those one million plus personalised journeys, start by way of laying a customer-driven foundation to place your self better inside the future.

How essential media publishers presently technique consumer experience (and monetization)

Let's study a few examples within the industry. HBO, domestic of Game of Thrones and Big Little Lies, gives on-line media streaming to paying subscribers:

HBO Example Subscription Portal

Whenever you click on a "Stream" call to movement (CTA), you'll see the popup above asking you to signal into your HBO Now or HBO Go account. If you're no longer already purchasing HBO, there's a CTA to "Get HBO" as properly.

You had this Zeitgeist display that also took place to attraction very plenty so to a digital-ahead technology of consumers at the identical time we, as a logo, had been going out and launching [a platform] so you can watch our content in a diffusion of different ways. Very fast, and very without difficulty, it have become this tentpole inside the digital space.

— Bernadette Aulestia, Executive VP of Global Distribution at HBO (thru Vulture)

That "Get HBO" CTA will deliver you to a pricing web page, which outlines the three distinct plans:

HBO Example Plans Page

This shape demonstrates a easy expertise of users' preferred effects. There's nobody-length-suits-all technique here. The three distinct plans goal wonderful audience segments with specific motivations, desires, pain factors and behavioral styles.

There is likewise a phase of the HBO web site that streams select unfastened media:

HBO-Example-Stream-Free-Media

This section is curated to lure visitors to start a loose trial, but additionally enables sharing. This demonstrates that initial shift from "one and accomplished" to "curated media hub". Sometimes the purpose is not absolutely to get a conversion (i.E. Cash) as quickly as possible. Often, it's approximately collecting behavioral data and positioning your self as a media hub, a destination.

This might be in which a dynamic paywall is brought primarily based on propensity modelling and behavioral technology. How probably is this viewer to begin a unfastened trial proper now? Should the CTA be proven now or 3 films from now? Should the CTA be for HBO Go or HBO Now? In mixture, this records can be used to tell the default adventure or better outline target market segments.

Fashion publication, Elle, takes a hybrid technique to media and monetization. You'll note paid marketing throughout the website online, along with inside character articles:

Elle example advert

Modern digital advertising allows for retargeting, of course. This serves media publications both within the experience that visitors have already verified cause to click and that it's a shape of personalization. If the goal is to hold visitors engaged and coming back on your booklet, every little little bit of customization counts.

Elle attempts to interact the traveller, to shift from a "one and done" interaction. In this article, there are some strategies at play:

User Engagement Example Elle

First, there's the reader Q&A format. Second, there's the choice to weigh in together with your very own opinion. You'll word that Elle has print subscription CTAs during the website online as properly:

Digital Ad Example for Print Issue of Elle

And unpaid digital subscription CTAs:

Unpaid Digital Subscription Ad Elle

Elle doesn't have a paid virtual subscription version like HBO, however that doesn't imply the book isn't beginning to adapt to the industry-huge digital transformation. There are many distinctive pricing fashions and methods to address this industry evolution, but they all must prioritize one component: an remarkable person experience.

Finally, take into account Medium, a textual content-based totally medium, like Elle, with a paid digital subscription version, like HBO:

Medium Sign Up Pop Up

Readers are given a positive variety of free Medium testimonies in line with month. If you join up for a unfastened account, you get an extra unfastened story.

Currently, it's 3 in line with month, however we are continually experimenting with special numbers of stories (and unique methods of supplying the meter) to understand how it influences how an awful lot human beings study, and our subscription business.

— Michael Sippey, VP of Product at Medium (through Medium)

Note that the sign on flows depend on Google and Facebook. While it's tough to say for certain, it's probably that this get admission to to supplementary private facts powers accelerated customization for readers.

Once you attain your article restrict for the month, you'll be hit with a mid-article CTA:

Mid Article Subscription Pop Up Example Medium

Like HBO, Medium is working to provide in advance fee to readers with a layer of loose content that piques hobby, before asking them to subscribe.

Medium is not unlike different digital media subscription corporations just like the Washington Post or The New Yorker—or maybe Spotify and Netflix. We sell content on a subscription basis. Like maximum paywalled websites, we supply a few stories away without spending a dime (currently, it's 3 in keeping with month). But in contrast to most paywalled publications, we depend totally on subscriptions (no advertising), and we've a mix of authentic and non-original content.

— Ev Williams, CEO at Medium (via Medium)

All of these corporations take a barely extraordinary method to monetization. None is inherently better than some other. What topics maximum is they've started out exploring the coronary heart of the virtual transformation and addressing the basis issue: figuring out and catering to their precise user's favored enjoy.

But that's most effective the beginning. From there, the aim ought to be to iterate and study through non-stop optimization.

Experimentation and optimization at The Motley Fool and The Wall Street Journal

The Motley Fool, a financial content material website online with a top class offerings version, recognizes that optimization is a full funnel affair. It's not pretty much acquisition and how many new customers you may get thru the proverbial door. It's additionally about how an awful lot those clients spend and how long they stay.

Across teams, we're all surely centered on purchaser lifetime value (LTV) for each degree. Whether it's a person's first product with us—getting them into a higher tier. And on the product facet, the greater we get people to renew, the higher their lifetime value. It's a win-win for both the purchaser and us: They get amazing stock recommendation and guidelines and we get to maintain them as a purchaser.

— Nate Wallingsford, Head of U.S. Marketing Operations & Optimization at The Motley Fool

They also prioritize learnings and insights over uplift. The energy of your optimization software, irrespective of your industry, is in how a lot you're mastering with every unmarried check. It's much less approximately what number of experiments you're jogging and how many of them are "prevailing" than maximum entrepreneurs consider.

For instance, The Motley Fool became experimenting with a well-known persuasion principle: social evidence. They ran numerous experiments. Each time, adding social proof reduced conversions, in spite of the fact that adding social proof is a broadly-familiar exceptional exercise.

Still, they continued with iterative exams to make sure that they had a clear know-how of how social proof become impacting their customers. After two experiments in which including factors of social proof had a terrible impact, Nate and the team determined to test eliminating factors of social proof. He explains:

"We ran some other experiment, this time at the order page—the stage of the funnel that includes the factor of purchase. In the version, all client testimónials [anóther fórm óf sócial próóf] had been remóved. Thís variatíón dóne terríbly, redúcing transactíóns, cómmón órder valúe and sales ín keepíng with sessión.

That was hónestly excítíng tó see. Even thóugh we had a decrease ín cónversíón rates acróss all 3 [sócial próóf] experíments, they generated this insíght that sócial próóf and testimónials are large ón the póint óf púrchase, hówever may alsó need tó be avóided ón the tóp óf the fúnnel."

This cónscióusness ón learnings and insights is specially crítical ín the media and leisúre enterpríse, ín which grówing perfórmance KPÍs (e.G. Sessión tó steer cónversíón charge and sales) is by nó means the móst effectíve púrpóse.

The Wall Street Jóurnal acknówledges hów experimentatíón ín the media enterpríse is a sensítíve stability amóng perfórmance and engagement.

It clearly depends ón what fórm óf check yóu're strólling and what the caúse óf yóur check ís. Fór a check ón óur newsletter center, we wóuld measúre úser engagement at the númber óf indivíduals whó súbscríbe fór newsletters ór túrn óut tó be tóggling ón their súbscríptión tó The 10-Póint (ór sómethíng públícatíón we míght be pushíng). Fór an artícle check, we ís próbably searching at úser engagement as qúantíty óf artícles accórdíng tó cónsúltatíón. What we stríve tó assúme ís, 'What ís the maxímúm crítical factór that we will fórce fór this particúlar píece óf actúal estate?' and then degree tówards that as óur predómínant KPÍ.

— Ólivía Símón, Óptimízatíón Manager at The Wall Street Jóurnal (thróugh Óptimízely)

Dígital transfórmatíón is nót sóme thing new tó The Wall Street Jóurnal. They've been tackling the evólutión head-ón fór pretty sóme tíme nów.

Ónce a media enterpríse has the prelíminary óptimízatíón basís ín lócatíón, typical dreams will in all likelíhóód begin tó exchange. It need tó end úp less abóut small adjústments and greater abóut cúltúral shifts, approxímately aútómating and develóping tó the next degree óf the dígital transfórmatíón:

As a lóng way as I'm cóncerned, the area óf experíence testing fór virtúal cómpanies is akín tó cóming acróss plútóníúm. It's júst an expónentíally móre pówerfúl manner óf creating búsíness selectíóns óver sóme thing, síncerely, withín the recórds óf mankínd. [...] This is the best way tó make cómmercial enterpríse selectíóns inside the fúture. This is hów every búsíness selectíón gets made góing fórward fór a virtúal próduct, anywhere feasíble.

— Peter Gray, VP of Product Optimization at The Wall Street Journal (through YouTube)

Taking on present day media with experimentation

To keep away from going the manner of the proverbial radio star, lean into the digital transformation that's sweeping the media and enjoyment industry. You don't necessarily want an costly tech stack or sophisticated information warehouse to get commenced. Laying a strong basis starts with a shift in thinking and a customer-centric approach:

Prioritize gaining a deep expertise of your audience's motivations, desires, ache factors and movements, frequently leveraging both facts and behavioral technology

Marry this knowledge with adventure mapping. Lay out the not unusual paths to conversion and audit them. What actions are you asking visitors to take? Are your appeals clear and distraction-unfastened? Is it difficult to take those moves? At which points do visitors currently "fall off" in these trips?

Using a strong optimization system, begin systematically experimenting with the ones journeys. At first, on-site. Then at the activation channel degree. Then layer in personalization and machine getting to know as you scale.

Streaming Drives Music and Video Revenue

While searching back on the beyond 12 months, one thing turns into crystal clear: streaming has grown massively in 2017 and could hold to make bigger rapidly in 2018 and past. In fact, streaming has become the number one sales driving force for each music and video, thereby greatly changing the manner people eat content and profoundly altering the manner wherein the enterprise and artists supply media to fanatics.

Digital song downloads are fast declining and being replaced by way of streaming, and it's clean to understand why that is going on. For the fee of a unmarried digital album down load, clients can gain get right of entry to to tens of hundreds of thousands of songs thru streaming offerings. And a developing variety of track enthusiasts are making this transfer, with over a hundred million people now subscribed to such offerings around the arena. According to the RIAA's 2017 mid-12 months record, the U.S. Recorded track enterprise skilled a 17% revenue growth inside the first half of this year in evaluation to 2016, and the primary driving force of that growth is streaming. While total digital revenue accounted for $three.2 billion, or 84% of the total industry value, an impressive sixty two% came from streaming, equaling $2.Five billion.

Video streaming revenue is also developing at a mind-blowing fee and not using a signs of slowing down. According to a report by Research and Markets, video streaming sales is forecast to increase from $30.29 billion in 2016 to an outstanding $70.05 billion via 2021, as consumers hold to embody pay TV and OTT solutions for streaming videos. Moving forward, it's expected that OTT will experience the most widespread market boom price primarily based on the increasing use of virtual platforms for the branding and advertising and marketing of products.

Among different things, streaming powers the increase of digital distribution, growing greater opportunities for impartial artists. The global of unbiased song distribution has additionally expanded significantly during the last year, as a growing wide variety of artists pass far from most important label deals to harness the personalised attention that indies can provide. UK-born corporation Ditto Music recently opened 12 new workplaces global to serve its flourishing roster, while virtual distribution and offerings enterprise FUGA has signed a deal with London-based Ignition Records, following its increasing partnerships with a large number of labels, which include Epitaph Records and Ultra Records. Even Google is entering into the track industry with a big investment in the new US-based totally digital music distribution startup UnitedMasters, which turned into based with the aid of the only-time President of Urban Music at Interscope Records, Steve Stoute.

Not especially, the opposition for streaming dollars has been developing exponentially. One of the most important latest tales turned into Disney's assertion that it is severing its distribution address Netflix in want of launching its own streaming service in 2019. And now it's far formally confirmed that Disney is shopping for some of 21st Century Fox's amusement assets, inclusive of the Fox film and television studio and a share of Hulu, so that you can permit Disney to boom its tv production to offer exceptional content material on its imminent streaming offering.

Although some are nevertheless skeptical approximately the lengthy-time inflow of streaming dollars, the streaming war is already fierce, and the conflict is certainly just beginning. As businesses try to secure dominance, technology will show to be the satisfactory weapon inside the streaming arsenal, thereby setting apart the leaders from the % in the coming year.

The Future Is Data-Driven, So What Does The Data Say?

Image for publish

As virtual and streaming answers more and more dominate the media enterprise, the function of facts is rapidly evolving from being crucial to turning into genuinely critical. At this point, streaming has already grown to be the primary sales driving force for each tune and video, making the powerful evaluation and usage of records an critical factor for fulfillment.

Technological improvements hold to convert each component of the media industry, from artist and concept development to licensing and the manner in which human beings consume their preferred music and TV indicates. And data sits inside the middle of this amusement revolution.

A thorough knowledge of song data is a crucial piece of the puzzle. Although this industry is notoriously unpredictable, facts's capability to as it should be expect and monitor insights makes it a very powerful device. An notable instance of this capacity turned into illustrated by way of Jón Davies, Director of EU Music Partnerships at Shazam, who said the app's wonderful potential to utilize statistics so that you can expect a success from six to 8 weeks in advance.

Data's significance in the evolving global of tv can't be overstated, as consumers are provided with more and more content and shipping choices pushed by means of facts-based decisions. An significant report from communications technology and services enterprise Ericsson genuinely exhibits the massive changes happening in TV. For one, the firm's study found that about 70% of consumers now watch tv and videos on a smartphone, which is double the percentage from simply five years ago. Ericsson predicts that by 2020, simply 10% of human beings will watch TV simplest on a conventional display, which would mark a 50% lower in assessment to 2010. And as increasingly more humans embody new technology, many experts believe that VR becomes an crucial factor of tv and video in the now not-too-remote future, with its social and immersive potential being realized in progressive and exciting new methods.

The skillful usage of facts is important for fulfillment, at the same time as mastering to be records-driven involves the extensive involvement of automation, gadget learning and AI. The complicated melding of first, 2d and third-birthday celebration facts lets in industry players to get the excellent effects out of an set of rules, thereby positioning facts to guide the way to creating more effective choices. Although people will continually be an crucial part of successful establishments, the complexities concerned in studying big portions of records should be treated by way of machines as we pass forward. Otherwise, the data will weigh down and confuse us in place óf preseñtiñg preciôús íñsíghts íñtó the cóñdúct and alternatíves óf the hastíly-rewórkíñg clíent landscape.

The Fútúre óf Creatíve Índústríes ín a Cónnected Wórld

Ímage fór súbmít

Technólógy ís dríving the vírtúal revólútíón even as cónsúmers íncreasíngly embrace revólútíónary answers fór their medía cónsúmptíón. Glóbal telephóne ównershíp cróssed the 2 bíllíón mark ín 2014 and ít's predícted tó attaín an extremely góód 4.6 bíllíón by means óf 2019. Óúr glóbal ís nów línked thrú the net and sócíal medía ín ways that exceed even the wíldest ímagínatíóns óf era ínnóvatórs fróm the past.

The develópíng get entry tó tó a myríad óf devíces, whích ínclúde smartphónes, smart speakers, cónnected aútómóbíles and línked hómes, allóws húmans tó be ón líne ín realíty everywhere and every tíme, cónsúmíng track and televísíón súggests ón-the-gó. And ít ís thís essentíal shíft ín cústómer behavíór and technólógícal advancement that's dríving the fast bóóm óf músíc and vídeó streamíng, alóng síde the grówíng sígnífícance óf ínfórmatíón and íts capabílíty tó adjúst the creatíón and shíppíng óf all medía kínds.

The cómíng yr will cónvey a large númber óf new útílízatíóns óf technólógy tó the leadíng edge. Ít's antícípated that vóíce-cóntrólled era will take ón a far large pósítíón ínsíde the clóse tó fútúre, fóllówíng the achíevement óf Amazón's Echó clever dómestíc devíce and the emplóyer's Alexa persónal assístant. Fórthcómíng ínnóvatíóns ín vóíce tech hóld the abílíty tó appreciably módify the manner whereín púrchasers lóók fór and cóncentrate tó músíc, resúltíng ín a góód larger effect ón sales transferríng fórward. Machíne masteríng and artífícíal íntellígence wíll an íncreasíng númber óf penetrate the medía índústry, becaúse the analytícal sectór develóps prógressíve sólútíóns tó make facts ínpút actíónable.

Perhaps móst sígníficantly, technólógy wíll preserve tó súbstantíally ímpróve the clíent experíence. Fróm ímpróvíng díscóvery abíltíes tó ímpartíng an íncreasíng númber óf persónalísed and cúrated cóntent delívery, the dígítal wórld hólds the capabílíty tó entertaín húmans ín súper appróaches that had been ónce relegated tó desíres.

☙

References

https://marketbusinessnews.com/economic-glossary/media-definition-meaning/

https://www.toppr.com/publications/civics/knowledge-media/what-is-media/

https://web sites.google.com/web page/multimediadreamwiki/1--introduction

https://en.wikipedia.org/wiki/Mass_media

https://open.lib.umn.edu/mediaandculture/chapter/1-three-the-evolution-of-media/

https://owlcation.com/humanities/A-Short-History-of-Media

https://www.nimcj.org/weblog-detail/timeline-of-the-evolution-of-mass-media.html

https://www.webpages.uidaho.edu/jamm445hart/timeline.htm

https://mprcenter.org/what-we-do/what-is-media-psychology/

https://bit.ly/3piYXLE

Https://www.admitkard.com/blog/2019/12/21/kinds-of-mass-media/

https://online.maryville.edu/blog/what-is-digital-media/

https://mooreks.co.united kingdom/insights/what-is-media-era/

https://bit.ly/2WHDLmd

https://bit.ly/3rt7wVZ

https://www.widerfunnel.com/blog/destiny-of-media-and-entertainment/

https://bit.ly/3heBf0h

About Author

Professor Sanjay Rout

Professor Doctor Sanjay Rout is an International Acclaimed Author, Scientist, Researcher, Futurologist, Knowledge Heuristic, Think-tank, and Policy Expert & Journalist. Honored as Global Best 50 Future Leaders in Innovation, Legal, Business & Future Technology by Thinkers-360, Eminent Researcher award by Green Thinker- Z , Best Scientist award by GECL International Foundation, Best Innovator award by MUGU International

ABOUT AUTHOR

Foundation, Author award by Story Mirror. He had received many National International Research Fellowships, Awards & honors for his work.

About Publisher

ISL PUBLICATIONS

ISL Publications is a global Research Development, Publication ,Advisory, Think-tank, Policy Research, Innovation Development, Business Consulting , Communication and Advisory Firm working on Future Business Solution.

www.ingramcontent.com/pod-product-compliance
Lightning Source LLC
Chambersburg PA
CBHW030844180526
45163CB00004B/1439